Contractor Logistics Support in the U.S. Air Force

T0146313

Michael Boito, Cynthia R. Cook, John C. Graser

Prepared for the United States Air Force

Approved for public release; distribution unlimited

PROJECT AIR FORCE

The research described in this report was sponsored by the United States Air Force under Contract FA7014-06-C-0001. Further information may be obtained from the Strategic Planning Division, Directorate of Plans, Hq USAF.

Library of Congress Cataloging-in-Publication Data

Boito, Michael, 1957–
    Contractor logistics support in the U.S. Air Force / Michael Boito, Cynthia R. Cook, John C. Graser.
        p. cm.
    Includes bibliographical references.
    ISBN 978-0-8330-4576-8 (pbk. : alk. paper)
    1. United States. Air Force—Weapons systems—Maintenance and repair.
    2. United States. Air Force—Weapons systems—Maintenance and repair—Costs.
    3. Logistics—Contracting out—United States. I. Cook, Cynthia R., 1965–
    II. Graser, John C. III. Title.

UG1103.B656 2009
358.4'1411—dc22

                                                                2009006827

The RAND Corporation is a nonprofit research organization providing objective analysis and effective solutions that address the challenges facing the public and private sectors around the world. RAND's publications do not necessarily reflect the opinions of its research clients and sponsors.

RAND® is a registered trademark.

Published 2009 by the RAND Corporation
1776 Main Street, P.O. Box 2138, Santa Monica, CA 90407-2138
1200 South Hayes Street, Arlington, VA 22202-5050
4570 Fifth Avenue, Suite 600, Pittsburgh, PA 15213-2665
RAND URL: http://www.rand.org/
To order RAND documents or to obtain additional information, contact Distribution Services: Telephone: (310) 451-7002;
Fax: (310) 451-6915; Email: order@rand.org

Preface

This monograph is part of a RAND Project AIR FORCE (PAF) project, "Weapon System Costing Umbrella Project." To improve the tools used to estimate the costs of future weapon systems, the project focuses on how recent technical, management, and government policy changes affect overall cost.

This monograph describes the increasing use of contractor logistics support (CLS) in the operating and support (O&S) phase of the weapon system life cycle and examines the associated funding, cost, performance, and management issues. Appendix B summarizes many of the laws, regulations, directives, and instructions that govern the use of CLS in the Air Force.

The research reported here was sponsored by Lt Gen Donald J. Hoffman, former Military Deputy, Office of the Assistant Secretary of the Air Force, Acquisition, and Blaise J. Durante, Deputy Assistant Secretary of the Air Force, Acquisition Integration, and was conducted within the Resource Management Program of PAF. The study's technical monitor is Jay Jordan, Technical Director for Cost and Economic Analysis Research of the Air Force Cost Analysis Agency.

This monograph should be of interest to those who plan for and manage the logistics support of Air Force weapon systems and to cost analysts who are responsible for O&S cost issues. A number of other PAF documents address weapon system acquisition issues and cost-estimating issues related to weapon system development and procurement. Recent PAF documents that address weapon system O&S include the following:

- In *Rethinking How the Air Force Views Sustainment Surge* (MG-372-AF), Cynthia R. Cook, John A. Ausink, and Charles Robert Roll, Jr., look at sustainment surge (increasing weapon system repair workload due to the operational demands of wartime or contingency operations) and how the nature of surge has changed, whether legislation has hindered management's ability to develop effective and efficient ways to manage surge, and whether it is possible to improve the effectiveness and efficiency of surge planning.
- In *Price-Based Acquisition: Issues and Challenges for the Defense Department Procurement of Weapon Systems* (MG-337-AF), Mark A. Lorell, John C. Graser, and Cynthia R. Cook assess price-based acquisition, a major acquisition reform the Department of Defense (DoD) is using in an effort to reduce costs and enhance acquisition efficiency. The essence of price-based acquisition is the notion that DoD should establish "fair and reasonable" prices for goods and services without extensive cost data from suppliers.
- In *Budget Estimating Relationships for Depot-level Reparables in the Air Force Flying Hour Program* (MG-355-AF), Gregory H. Hildebrandt develops estimating models to explain the historical net sales of flying depot-level reparables. The models relate net sales to aircraft characteristics, operational tempo, and time-related variables.

RAND Project AIR FORCE

RAND Project AIR FORCE (PAF), a division of the RAND Corporation, is the U.S. Air Force's federally funded research and development center for studies and analyses. PAF provides the Air Force with independent analyses of policy alternatives affecting the development, employment, combat readiness, and support of current and future aerospace forces. Research is conducted in four programs: Force Modernization and Employment; Manpower, Personnel, and Training; Resource Management; and Strategy and Doctrine.

Additional information about PAF is available on our Web site: http://www.rand.org/paf/

Contents

Preface iii
Figures ix
Tables xi
Summary xiii
Acknowledgments xix
Abbreviations xxiii

CHAPTER ONE
Introduction 1
Maintenance of Weapon Systems and the Use of Contractor
        Logistics Support. 1
Purpose of This Monograph 4
Research Approach. 5
Organization of This Monograph 7

CHAPTER TWO
Background and Policy Guidance on CLS Use 9
Background 10
Key Laws Affecting CLS Use. 10
Key DoD Directives and Instructions That Affect CLS 12
Key Air Force Direction That Affects CLS 13
Implications of Laws and Regulations That Affect CLS Use 15

CHAPTER THREE

Funding and Cost Issues Associated with CLS Use 17
Findings and Observations: CLS Funding and Costs 18
   How Does the Air Force Define CLS for Funding and Costing? 18
   What Programs Use CLS, and How Much Does Each Spend
      on CLS? 24
   Programs That Account for the Growth in CLS Over the Last
      Seven Years 26
   What Are the Key Cost Drivers for CLS? 29
Observations and Conclusions About CLS Funding and Costs 33

CHAPTER FOUR

Assessing the Performance of CLS 35
Cost and Performance: Comparing CLS with Organic Support 35
   How Does the Government Measure CLS Performance and
      Using What Metrics? 35
   What Has Contractor Cost Performance Been Like for Major
      Programs Using CLS? 38
   How Effective Has CLS Proven for Major Programs Using CLS? 43
   Discussion on CLS Supply Support 45
   How Does CLS Price Growth Compare to Organic Cost Growth? 47
   Do Contractors Have Inherent Advantages or Disadvantages in
      Performing Some Tasks? 49
Observations on CLS Performance 51
Conclusion 52

CHAPTER FIVE

CLS Management 55
Current Processes for Choosing CLS and Ongoing CLS
      Management 55
Reasons Existing Programs Use CLS 57
How Are Tasks on CLS Contracts Defined and Funded? 65
   Why Do CLS Contracts Have So Little Variable Funding? 68
   How Much Insight Does the Government Have Into the
      Contractor's Costs? 69

How Are CLS Contract Prices Determined? 72
Observations on CLS Management 73

CHAPTER SIX
Implications for Cost Analysts 75
Funding Sources May Shift at Different Stages of Support................. 75
CLS Affects the Amounts and Proportions of Costs Reported in
    Non-CLS O&S Elements... 76
The Nature and Scope of CLS Tasks Differ Among Programs.............. 76
Cost and Cost Growth 77
Some CLS Costs Are Accounted for Differently Than Are the
    Corresponding Organic Costs.. 78
It Is Difficult to Generate Cost-Estimating Relationships for Total
    System O&S Costs Because Funding Constraints Affect Much
    of the Total Cost 78

CHAPTER SEVEN
Summary and Recommendations 81
Summary of Findings... 81
Recommendations 82
    Require Centralized Decisions on Buying Design and Technical
        Data or Use Rights to Data.. 83
    Require a Uniform Format for Cost Data 85
    Provide Centralized Guidance to Achieve Flexibility 87
    Strengthen Centralized Expertise to Optimize CLS Use 88
    Retain Choices for Logistics Services..................................... 91

APPENDIXES
A. **Comparison of Supply-System Performance on CLS and Organic
    Programs** 93
B. **Laws, Directives, Regulations, Instructions, and Reports That
    Affect CLS Use** 99

References 125

Figures

1.1.   Air Force Spending on CLS and Organic Maintenance 3
3.1.   Air Force CSS for Weapon Systems 19
3.2.   Air Force CLS Spending by Type of System, FY 2006 22
3.3.   Air Force Aircraft Operating and Support Costs,
        FY 2006 23
3.4.   Top 14 Largest Air Force Aircraft CLS Programs in
        FY 2006 and Their Growth Since FY 2000 24
3.5.   Growth in Aircraft CLS and Aircraft Organic
        Maintenance Funding 26
3.6.   Aircraft CLS Program Costs, Largest Cost-Growth
        Programs 27
3.7.   Inventory of Aircraft with Growing CLS Programs 29
3.8.   Key Drivers of Aircraft CLS by Element of O&S Cost 32
5.1.   CLS Use and Fleet Size 63
A.1.   TNMCS Rates and Standards, Selected Trainer Aircraft 94
A.2.   TNMCS Rates and Standards, Selected Cargo Aircraft 95
A.3.   TNMCS Rates and Standards, Selected Fighter Aircraft 96
A.4.   TNMCS Rates and Standards, Tanker Aircraft 97

Tables

3.1. Fourteen Largest Air Force Aircraft CLS Programs............25
4.1. Metrics Used to Assess CLS Performance 37
5.1. Reasons Cited for Use of CLS....................................61
5.2. Guaranteed and Variable Funding Amounts in the Nine
      Largest CLS Contracts...67

Summary

The Air Force devotes enormous resources to operating and maintaining its weapon systems. In fiscal year (FY) 2006, the Air Force spent almost $36 billion on weapon system O&S, measured in constant FY 2004 dollars.[1] The Air Force has a range of choices when considering how best to sustain weapon systems and components. It can do the work in-house using organic facilities, it can pay contractors to do the work (subject to some congressionally imposed limits), or it can engage in a mix of the two approaches.[2]

This monograph addresses CLS, which is defined as contractor sustainment of a weapon system that is intended to cover the total life cycle of the weapon system and generally includes multiple sustainment elements. CLS does not include interim contractor support, a temporary measure for a system's initial period of operation before a permanent form of support is in place. CLS also excludes contractor sustainment support for a specific sustainment task that the Air Force would otherwise conduct itself; a typical example would be a weapon system's prime contractor providing sustaining engineering.[3]

---

[1] O&S includes all costs of operating, maintaining, and supporting a fielded system, including costs for personnel; consumable and repairable items; organizational, intermediate, and depot maintenance; facilities; and sustaining investment. Data from an Air Force Total Ownership Cost management information system query in January 2007.

[2] Federal laws require that government facilities conduct at least one-half of all depot maintenance work and that the government retain certain core maintenance capabilities. Chapter Two discusses this and provides specific references.

[3] In practice, there is some overlap among the various kinds of contract support. The Air Force identifies and funds the varieties of contract sustainment support using element-of-

The Air Force has increasingly chosen CLS as an alternative to organic support of weapon systems over the last several years. The Air Force increased its use of CLS by more than 150 percent in constant dollars from FY 2000 to FY 2006, a rate far greater than the 30 percent increase in spending on weapon system O&S over the same period.

Despite the Air Force's increased use of CLS, several of the unanswered questions about its management and use might be of interest to decisionmakers. We examine these questions, and when appropriate, provide recommendations for more effective use of CLS:

- What is driving the growth of CLS in the Air Force?
- How has contractor performance under CLS compared to initial estimates of cost and performance?
- What are the key cost drivers for CLS?
- How are the prices for CLS contracts determined?
- Do weapon systems have characteristics that are associated with using CLS; if so, what are they?
- How does the Air Force manage its compliance with laws governing the use of CLS?
- Does using CLS have disadvantages?
- What does using CLS imply for O&S cost estimating?

We approached these questions in four ways. First, we reviewed the laws, regulations, and instructions that govern the use of CLS in DoD and especially in the Air Force. This helped us understand limits and requirements that Congress and DoD have imposed on the Air Force for the use of CLS, as well as the official implementation of policies and procedures. We also reviewed reports from the Government Accountability Office (GAO) and DoD Inspector General, which were helpful in understanding problems and issues with the use of CLS over time.

---

expense investment codes (EEICs) in its financial system. EEIC 578 is intended to capture CLS. We have used funds coded as EEIC 578 to identify CLS in this study. Chapter Three describes this coding more completely. Air Force Instruction (AFI) 63-111 dated October 21, 2005 defines the kinds of contract support.

Second, we examined the Air Force Total Ownership Cost management information system to understand the magnitude of CLS use and cost trends for the Air Force as a whole, by type of system, and by individual program.

Third, because the Air Force does not require detailed or uniform financial reporting of CLS costs in a corporatewide financial system, we obtained CLS brochures for individual programs to gain greater insight into the nature of CLS. Programs that do use CLS prepare these brochures annually. Each brochure contains a narrative describing the support provided and its cost by task. These brochures allowed us to address questions about key CLS cost drivers—specifically, about the kinds of tasks for which CLS is used.

Fourth, we spoke with a wide variety of people knowledgeable about the use of CLS. We held discussions with representatives of roughly a dozen large programs that use CLS. We also interviewed personnel experienced in logistics, contracting, and CLS financial and program management. These people worked at Air Force operational commands and Air Force Materiel Command, Headquarters U.S. Air Force (HQ USAF), the Office of the Secretary of Defense, the GAO, and the weapon system contractors that provide CLS.

This combination of approaches allowed us to address most of the research questions satisfactorily. Some issues regarding CLS performance remained unresolved, as we will discuss.

We found that 86 percent of the Air Force's CLS spending is on aircraft systems, with the remainder on space, missile, munitions, and other kinds of systems. (See p. 22.) Most of the growth in CLS spending in this decade has been in aircraft systems, with the C-17 and F-22 programs increasing the most because more aircraft are being delivered and because both programs recently transitioned from interim contract support to CLS. We found no evidence that the costs of ongoing and long-term CLS contracts are increasing at a faster rate than comparable organically supported programs. Rather, the increase in CLS spending is due mostly to decisions to support most new aircraft systems with CLS, while the legacy systems that they replace tend to be supported organically. (See pp. 24–29.)

We were unable to determine how CLS had performed relative to initial estimates because we found that initial estimates generally either were not developed or were not documented and retained. (See pp. 40–41.)

The key cost drivers for CLS on aircraft programs are depot maintenance for airframes and engines and the repair and replacement of parts. (See pp. 31–32.) Aircraft CLS programs have a wide scope of tasks, so contractors on some programs provide a substantial number of field-support representatives, or technicians, who perform maintenance at the flight line. Contractors on other programs provide much of the sustaining engineering. On practically all, if not all, aircraft CLS programs, contractors provided supply-chain management. (See pp. 20–21, 31–32, 93.)

We found that CLS prices for major weapon systems are seldom determined by competition. Exceptions to this norm are typically for commercial-derivative products. (See p. 72.) Prices depend on the type of contract. CLS contracts use a variety of contract vehicles and types. For cost-type contracts or tasks, the contractor is reimbursed for costs incurred plus a fee. For fixed-price contracts or tasks, government personnel generally examine the number of labor hours and the material costs the contractor has proposed to determine whether both are reasonable. The price is determined by the negotiated labor hours, the contractor's hourly labor rate(s), expected material usage and cost, plus a fee (profit). (See pp. 72–74.)

CLS contracts often guarantee a large amount of funding to the contractor in each fiscal year. This limits the flexibility of the Air Force to reduce funding levels without violating the terms of the contract. (See pp. 68–69.)

Competition is often impossible because the government lacks the technical data or the data rights needed to allow third parties to maintain the equipment, so only the original equipment manufacturer, which has the technical data, can do the maintenance. (See p. 58.)

We found that the availability of CLS cost and performance data varied among individual program offices. (See pp. 69–74.)

Several weapon-system characteristics were associated with the use of CLS, including programs that were commercial derivatives, were highly classified, were complex, had a small fleet, or had started as advanced concept technology demonstrations. Two additional conditions associated with the use of CLS, although not with characteristics of the weapon systems, were a lack of data rights and the decision of a senior Air Force official. (See pp. 57–65.)

Certain limitations affected our ability to address all the issues satisfactorily. The most serious is the lack of detailed cost and performance data on CLS contracts, which limits our ability to assess the cost and performance of CLS relative to initial estimates or government performance of comparable work. The lack of detailed cost and performance data on CLS contracts also severely limited our ability to provide improved tools or guidance to cost estimators.

Also note that we did not specifically address issues associated with the use of performance-based logistics or public-private partnerships, which are sustainment approaches DoD has emphasized for the last several years.

The final chapter discusses five changes that should improve the Air Force's ability to use CLS effectively. The Air Force is in the process of implementing some of the changes:

1. To preserve the option of sustainment by organizations other than the contractors that manufactured the equipment, the Air Force should require centralized decisions on buying design and technical data or usage rights to such data. (See pp. 83–85.)

2. To facilitate future analysis and estimation of O&S costs, the Air Force should require collection of CLS cost data in a standardized format, as specified by the Office of the Secretary of Defense Cost Analysis Improvement Group, and should retain the data centrally. (See pp. 85–87.)

3. To ensure that the corporate Air Force has the flexibility to adjust funding levels for all aircraft sustainment programs, the Air Force should provide centralized guidance to achieve flexibility in CLS contracts. (See pp. 87–88.)

4. To improve its ability to manage CLS across the enterprise, the Air Force should strengthen data collection and analysis and expertise and make the data and expertise available to pro-

gram office personnel. This could be done by centralizing and strengthening an organization with logistics responsibilities and/or by strengthening a career field, such as acquisition logisticians. (See pp. 88–91.)

5. The Air Force should strive to retain choices for logistics services over the life cycle of the weapon system. The first four changes support this goal. (See pp. 91–92.)

Acknowledgments

We would like to thank Lt Gen Donald J. Hoffman, Military Deputy, Office of the Assistant Secretary of the Air Force, Acquisition; Blaise J. Durante, Office of the Deputy Assistant Secretary of the Air Force, Acquisition Integration; and Maj Gen Art Morrill, HQ USAF/ILP,[4] for cosponsoring this research, and Jay Jordan, Air Force Cost Analysis Agency, for providing technical direction. Jay Jordan also provided useful comments on a draft of the monograph.

We also wish to thank the two leaders of the CLS Integrated Product Team (IPT) formed by HQ USAF Logistics, Installations, and Mission Support, Lt Col Randy Mauldin and Yvonne Fernandez, who graciously allowed us to join the IPT and supported our study efforts. Our membership in the IPT allowed us to study and participate in the solution of issues along with the rest of the team. Membership on the IPT also made it easier to identify and communicate with people in the Air Force who are knowledgeable about CLS.

Many people in the Air Force, Office of the Secretary of Defense, and other organizations were generous in sharing their time and knowledge of logistics support with us. In particular, we would like to thank Grover Dunn, HQ USAF/A4I, for his time and insights.

In many cases, we attended meetings or participated in teleconferences with several representatives from an office. Rather than listing the names of all of the people with whom we spoke, we list their organizations in alphabetical order:

---

[4] We show the rank and organizational affiliations of the individuals acknowledged here at the time we worked with them on this study. Some individuals have since moved to other organizations.

- Air Force Materiel Command, Acquisition Logistics Division
- Air Force Materiel Command, Depot Programs Division
- CLS providers, various representatives
- F-35 Program Office
- GAO
- HQ USAF, Logistics, Installations, and Mission Support and Maintenance
- HQ USAF, Logistics, Installations, and Mission Support
- Operational commands, various representatives
- Office of the Secretary of Defense, Cost Analysis Improvement Group
- Under Secretary of Defense for Acquisition, Technology, and Logistics
- Program offices, various representatives: C-17, Distributed Common Ground Systems , F-22, F-117, Joint Surveillance Target Attack Radar System, RC-135, Space-Based Infrared Systems, T-1, T-6, T-38, U-2
- Secretary of the Air Force, Acquisition Center of Excellence Rosslyn, Virginia
- Office of the Deputy Assistant Secretary of the Air Force, Contracting
- Office of the Deputy Assistant Secretary of the Air Force, Acquisition Integration
- Office of the Deputy Assistant Secretary of the Air Force, Cost and Economics
- Office of the Deputy Assistant Secretary of the Air Force, Installations, Environment, and Logistics
- Warner Robbins Air Logistics Center

We also wish to thank our RAND colleagues Nancy Moore and Ray Pyles, who shared insights on logistics issues, and Obaid Younossi and Laura Baldwin for reviewing drafts of this monograph and providing useful feedback that improved its quality. Mary Chenoweth and Ellen Pint conducted thorough reviews and provided comments and suggestions that provided further improvements. Charles Robert (Bob)

Roll, Jr., provided the initial guidance on the project and support during the research. He passed away before the project was completed.

Abbreviations

ACTD	advanced concept technology demonstration
AFB	Air Force base
AFEE	Air Force Element of Expense
AFI	Air Force Instruction
AFMC	Air Force Materiel Command
AFPD	Air Force Policy Directive
AFSO 21	Air Force Smart Operations for the 21st Century
AFTOC	Air Force Total Ownership Cost
ALC	air logistics center
BRAC	Base Realignment and Closure
CAIG	Cost Analysis Improvement Group
CLS	contractor logistics support
COTS	commercial-off-the-shelf
CSS	contract sustainment support
DLR	depot-level reparable
DoD	Department of Defense
DoDI	Department of Defense Instruction
EEIC	element of expense investment code
FAR	Federal Acquisition Regulation
FSR	field support representative

FY	fiscal year
GAO	Government Accountability Office
HQ	headquarters
HQ USAF/A4/7	Headquarters U.S. Air Force, Logistics, Installations, and Mission Support (Formerly HQ USAF/IL and HQ USAF/LG)
HQ USAF/A4I	Headquarters U.S. Air Force, Logistics, Installations, and Mission Support, Innovation and Transformation
HQ USAF/ILP	Headquarters U.S. Air Force, Installations and Logistics, Plans and Resources
ICS	interim contract support
IG	inspector general
IOC	initial operational capability
IPT	integrated product team
LCMP	life-cycle management plan
MERLIN	Multi-Echelon Resource and Logistics Information Network
O&M	operation and maintenance
O&S	operating and support
OEM	original equipment manufacturer
OLAP	Online Analytic Processing
OMB	Office of Management and Budget
OSD	Office of the Secretary of Defense
PBL	performance-based logistics
PDM	programmed depot maintenance
PEO	program executive officer
POS	preoperational support
SAF/AQ	Assistant Secretary of the Air Force, Acquisition
SORAP	source of repair assignment process

TNMCS	total not mission capable supply
UAV	unmanned aerial vehicle
USC	United States Code

CHAPTER ONE

Introduction

Maintenance of Weapon Systems and the Use of Contractor Logistics Support

The Air Force devotes enormous resources to operating and maintaining its weapon systems. In fiscal year (FY) 2006, the Air Force spent almost $36 billion on weapon system operating and support (O&S), measured in constant FY 2004 dollars.[1] The Air Force has a range of choices when considering how best to sustain weapon systems and components. It can do the work in house, using organic facilities; it can pay contractors to do the work (subject to some congressional limits); or it can apply a mix of the two approaches.[2]

Organic repair is a massive undertaking. Work takes place at government depots, such as the three major air logistics centers (ALCs), a number of intermediate repair facilities, and at flight lines on operating bases around the world. Managing organic repair efficiently has been a challenge for the Air Force throughout its existence. The Department of Defense (DoD) reported that the Air Force spent $5.1 billion on organic depot maintenance and an additional $4.6 billion for private

---

[1] This number includes direct costs, such as crew, maintenance personnel, fuel, and repair parts, and indirect costs, such as base operating support costs (Air Force Total Ownership Cost [AFTOC] management information system, January 2007).

[2] Federal laws require that at least one-half of depot maintenance work be performed at government facilities and that the government retain certain core maintenance capabilities. Chapter Two discusses these requirements in greater detail.

depot maintenance in FY 2005 in then-year dollars (Office of the Secretary of Defense [OSD], April 2006).

The Air Force also purchases sustainment services from a large range of commercial companies. The term *contract support* includes various types of support for systems, equipment, and end items that contractors provide. Contract sustainment support (CSS) provides for one or many logistics tasks or sustainment elements that organic logistics organizations would otherwise perform and can provide for all or part of the weapon system. Contractor logistics support (CLS) is a subset of CSS; the key distinctions are that CLS normally involves *multiple* sustainment tasks, usually *for the life* of the weapon system. Examples of common CLS tasks are aircraft and engine overhaul, repair and replenishment of parts, sustaining engineering, and supply chain management.

Both organic sustainment and CLS have their benefits. Organic sustainment gives the Air Force maximum control over when and how the work gets done. The Air Force benefits directly from any efficiency improvements. The Air Force also has a guaranteed source of supply.[3]

Using contractors for logistics support also has strengths. Proponents of CLS contend that profit-seeking companies will bid for the work at a lower cost, with competition as the spur, and will do the work more efficiently to earn a profit. This is particularly true if the system is a derivative of a commercial product and if an existing group of companies competes for the work (as for the KC-10). Having the prime contractor, the original equipment manufacturer (OEM), perform the repair work could offer economies of scale because this organization would manage the total supply chain. An OEM providing CLS can also offer savings by using existing repair facilities and specialized tooling, because production tooling can later be repurposed for maintenance and repair.

Both sides of the debate offer reasoned arguments but little documentation to tip the balance of evidence in favor of either organic

---

[3] The congressional debate on these benefits resulted in the so-called *50-50 rule*, which addresses keeping capacity in house as a way to reduce risks in case of a surge or for other reasons. See the discussion in Cook et al., 2005.

or CLS. We do know that the Air Force has often chosen CLS over organic support on major programs during the past several years. The Air Force's use of CLS more than doubled in constant dollars from FYs 2000 to 2006, a rate of increase far greater than for its spending on organic maintenance of weapon systems.[4] Figure 1.1 illustrates this growth.

O&S costs for the Air Force jumped markedly in FYs 2002 and 2003 because of wartime operations, and the increase in CLS costs mirrors the overall Air Force trend over the period. However, the additional

Figure 1.1
Air Force Spending on CLS and Organic Maintenance

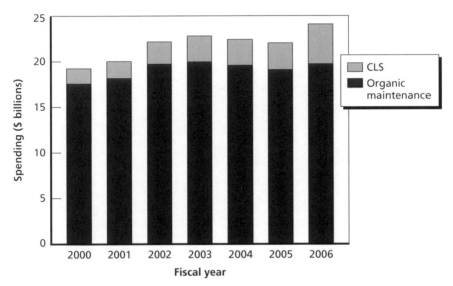

SOURCE: AFTOC OLAP report by weapon system and CAIG element, May 2007.
NOTE: Constant FY 2004 dollars. Organic maintenance costs are approximated as total operation and support costs less crew, fuel, CLS, and indirect support.
RAND MG779-1.1

---

[4] We approximated organic maintenance costs by subtracting the costs of crews, fuel, CLS, and indirect support from weapon system O&S costs. These are maintenance costs directly associated with weapon systems and do not include indirect O&S costs associated with a base, site, or location. The organic weapon system maintenance costs have grown 9 percent in constant dollars during the same period. (AFTOC Online Analytic Processing [OLAP] data by weapon system and by cost analysis improvement group [CAIG] element, May 2007.)

large increase in CLS costs in FY 2006 contrasts with O&S spending for organic logistics support. The increase of more than $1 billion in one year in CLS costs reflects primarily the transition of the C-17 and F-22 programs from interim contract support (ICS) funded in the aircraft procurement account to CLS. CLS costs for these programs are expected to grow as their inventories increase.

Purpose of This Monograph

The Air Force is facing severe budget pressures as it tries to simultaneously recapitalize its aging fleet of aircraft, operate and maintain them in support of the global war on terrorism, and accomplish other missions. Responses to these budget pressures include a reduction of manpower authorizations and a servicewide initiative of business process improvements known as Air Force Smart Operations for the 21st Century (AFSO 21).[5] In this environment of limited resources and competing demands for those resources, it is especially important for decisionmakers to be able to allocate resources wisely.

This monograph has two objectives. One objective of this research was to understand CLS use in the Air Force better and make recommendations, if called for, to enhance decisionmaking in this area. We wanted to determine what data exist to make comparisons of cost and performance and inform decisions and how the Air Force might retain options once a logistics support provider is chosen.

A secondary goal was to highlight challenges for cost estimators who are working with the O&S costs of Air Force programs and provide better estimating tools, if possible. We were unable to develop cost-estimating tools because CLS costs are not collected and reported in a standard and detailed format suitable for cost analysis.

Understanding CLS costs is important for several reasons. A better understanding of CLS costs would allow policymakers and cost

---

[5] AFSO 21 encompasses the implementation of lean initiatives that the Secretary of the Air Force and the Chief of Staff of the Air Force directed in a November 7, 2005, memorandum.

estimators to compare CLS costs with organic support costs by task, understand what is driving costs for CLS contracts and thereby forecast potential growth in costs better, determine whether incentives are working to reduce costs; determine whether contract profits are reasonable, and include CLS costs in cost models along with organic support costs.

Research Approach

We addressed Air Force use of CLS in four ways. First, we reviewed the laws, regulations, and instructions that govern CLS use in DoD, especially in the Air Force. This helped us understand the limits and requirements Congress and DoD have imposed on Air Force use of CLS, as well as the official implementation of policies and procedures. We also reviewed relevant Government Accountability Office (GAO) and DoD Inspector General (IG) reports, which helped us understand problems and issues related to current and historical use of CLS. Appendix B summarizes this literature review.

Second, we examined available Air Force financial system data to understand the financial magnitude of CLS use and to observe cost trends, for the Air Force as a whole, by type of system and by individual program. The AFTOC management information system was our main source of financial information. AFTOC is the authoritative source because it provides cost information on all major Air Force weapon systems, including aircraft, space, and missile systems, in a standard format. These data allowed us to corroborate and quantify in dollars what we learned in the other approaches about the kinds of systems for which CLS is used and where CLS use is growing. Unfortunately, because Air Force policy does not require detailed or uniform financial reporting of CLS costs, AFTOC does not contain detailed CLS costs, and generally displays all CLS costs as a lump sum for each weapon system.

Third, we obtained CLS brochures for individual programs for greater insight into the nature of the CLS and its cost. Each program office using CLS prepares an annual CLS brochure that includes a narrative describing and costing each support task for the system, ultimately providing greater detail than AFTOC supplies.[6] CLS brochures contain CLS cost estimates for the fiscal year in which they are submitted and for future years. The numbers in the brochures do not reflect actual obligations and may thus differ from the actual obligations reported in AFTOC, although the differences are usually fairly small for the fiscal year in which the brochures are prepared. Although the brochures reflect estimates rather than actual CLS expenditures, they are useful because of their level of detail. The brochures allowed us to address questions about the key cost drivers of CLS, that is, the kinds of tasks for which CLS is used.

Fourth, we spoke with a wide variety of people knowledgeable about CLS use. We joined an ongoing CLS Integrated Product Team (IPT) organized by Headquarters U.S. Air Force, Logistics, Installations, and Mission Support (HQ USAF/A4/7), which proved extremely helpful. The IPT was formed to assess the risks associated with funding reductions and to determine mechanisms for establishing flexibility and standardization in future CLS contracts and for making appropriate risk trade-offs. Our participation in the IPT made us aware of the visibility and flexibility issues, and gave us access to program office personnel who had extensive experience with CLS on a variety of programs. We also spoke individually with IPT participants representing roughly a dozen large programs using CLS.

We also established relationships with personnel experienced in logistics, contracting, and CLS financial and program management from Air Force operational commands, Air Force Materiel Command (AFMC), Headquarters U.S. Air Force (HQ USAF), OSD, GAO, and weapon system contractors providing CLS. By holding only not-for-attribution discussions with these individuals, we have ensured that the findings reported here are not associated with any individual or office.

This combination of approaches allowed us to address most CLS use issues satisfactorily, although some issues about CLS performance remain unresolved.

---

[6]   The level of reporting detail in the brochures varies from program to program and depends upon the cost reporting negotiated on each CLS contract.

Organization of This Monograph

Chapter Two provides the background for the Air Force's use of CLS. The discussion includes a review of applicable laws, DoD and Air Force instructions and regulations, and GAO and DoD IG reports and a description of the current Air Force process for choosing a source of repair for Air Force weapon systems.[7] Chapter Three summarizes Air Force funding of CLS and related contract sustainment services; breaks out CLS by type of weapon system; and most closely, examines CLS costs on aircraft programs. The material on aircraft programs includes an assessment of the kinds of tasks typically performed on CLS contracts. In Chapter Four, we discuss using cost and effectiveness for assessing CLS performance, and in Chapter Five, we address the Air Force's process for choosing CLS and the ongoing CLS management and associated issues. Chapter Six covers the implications of the findings for cost estimators addressing Air Force O&S costs. In Chapter Seven, we summarize the overall discussion and recommend policy changes. Appendix A compares total not mission capable–supply rates and standards for selected CLS and organically supported aircraft. Finally, Appendix B summarizes the relevant laws, reports, and DoD and Air Force directives, regulations, and instructions.

---

[7]   Appendix B provides further detail on these references.

CHAPTER TWO

Background and Policy Guidance on CLS Use

This chapter summarizes the results of our review of the legislative and policy history of the Air Force depot system and Air Force use of CLS. Some of the challenges the Air Force currently faces in using and managing CLS spring from existing law. Two legal requirements in particular are relevant to this study.

First, 10 USC 2466 requires doing at least 50 percent of depot maintenance in house, using organic government organizations. Therefore, at most, contractors can perform only the remaining 50 percent.

Second, Title 10 of the United States Code, Section 2464 (10 USC 2464) requires DoD to determine which logistics capabilities are considered *core*; these must be owned and operated by the government. *Core logistics capabilities* are defined as the technical competence and resources necessary to maintain and repair weapon systems during peacetime and during mobilizations, contingency situations, and other national emergencies. A related section of the code, 10 USC 2462, requires DoD to procure noncore maintenance from private sources if they are cheaper than organic sources.

This chapter also summarizes key DoD and Air Force directives, regulations, and instructions that implement legislation and otherwise affect Air Force use of CLS and other forms of contract support. The review in this chapter is not intended to be exhaustive but rather to summarize key legislative and administrative direction that affects the Air Force's use of CLS. Appendix B offers a broader summary.

Background

The current Air Force organic depot system had its roots in activities leading up to and during World War II, when the U.S. armed forces expanded rapidly in response to the aggression of the Axis powers. Aside from Kelly Air Force Base (AFB), which had been an Army Air Corps base since before World War I, all the other depots were opened in the late 1930s or early 1940s. Because the private sector was fully engaged in producing new equipment for the armed forces, these public-sector depots were established and expanded to perform maintenance on weapon systems (Heivilin, 1993). After World War II, the Cold War established the need for permanent depot maintenance expertise to repair increasingly complex weapon systems.

Despite cycles of downsizing and expansion during the last half of the 20th century in response to Korea, Vietnam, and Operation Desert Storm, the Air Force maintained its five major depots until the Base Realignment and Closure (BRAC) Commission in 1995 recommended closure of two ALCs.[1] Kelly and McClellan AFBs closed in 2001, leaving the Air Force with Warner-Robins, Hill, and Tinker AFBs as sites for its major depot maintenance activities.

Key Laws Affecting CLS Use

Over the years, Congress has maintained an active interest in the location of organic depots. This interest stems not only from a national security perspective, the desire to ensure that national defense had strong support from the depot-level maintenance and repair activities, but also from more parochial reasons. Congress formally recognized the requirement for a "core logistics activity" for the first time in the 1984 National Defense Authorization Act, which directed DoD to maintain an in-house logistics capability that could respond to mobi-

---

[1] The Air Force also closed a sixth depot at Newark AFB, Ohio, which specialized in the repair of guidance and navigation systems, in 1996.

lizations, national contingencies, and other emergency requirements (Cook, Ausink, and Roll, 2005).[2]

This requirement is found in 10 USC 2464, "Core Logistics Capabilities." This section, originally passed by the 98th Congress in 1984, discusses the need for core logistics capabilities that are government owned and operated (including government personnel and equipment), directs the Secretary of Defense to identify core logistics capabilities, and defines core logistics capabilities as those that are necessary to maintain and repair the weapon systems and other military equipment, among other provisions.

Congress has also taken an active role by directing what proportion of DoD depot maintenance workload could be performed by contractors (private sources) and what must be performed organically, at DoD depots. One reason for this directive is the large number of federal jobs involved with the depot activities in several states. In fact, Air Force depots are major employers in all states in which they are located. Supporters of the depots have formed the House Military Depot and Industrial Facilities Caucus, a loose confederation of members of the House of Representatives interested in DoD's depots and their activities. To codify their interests, Congress has passed legislation to guide depot-level maintenance and repair activities, limit using contractors in providing that support, and require regular reporting on spending in this area.

The key legislation is found in 10 USC 2466, "Limitations on the Performance of Depot-Level Maintenance of Materiel." This section discusses limitations on the amount of depot-level maintenance and repair workload that contractors can perform relative to the amount government facilities perform. The original restriction on contractor workload was a maximum of 40 percent, as set in 1988. Congress increased that maximum to 50 percent in the FY 1998 Defense Authorization Act. The 2005 version of 10 USC 2466 sets the current limit:

> Not more than 50 percent of the funds made available in a fiscal year to a military department or a Defense Agency for depot-level

---

[2] See Chapter Two of the 2005 report for a further discussion of the history of congressional actions related to this issue.

> maintenance and repair workload may be used to contract for the performance by non-Federal Government personnel of such workload for the military department or the Defense Agency.

However, in the same legislation, Congress also changed its definition of *workload*, specifying that DoD also include work subcontracted to private companies via public depots and include ICS, which is funded differently from CLS and which had previously been excluded.

A third important provision is 10 USC 2462, "Contracting for Certain Supplies and Services Required When Cost Is Lower." This section directs the Secretary of Defense to procure each supply or service necessary to accomplish the authorized functions (other than those the Secretary of Defense determines military or government personnel must perform) from a source in the private sector if that source can provide the supply or service at a cost lower than it would cost DoD to provide the supply or service. This legislation implies that the DoD will determine the low-cost provider and award the work accordingly.

The National Defense Authorization Act for FY 2007 included a provision that should affect the ability of program managers to compete sources of repair over the long term. Public Law 109-364, Section 802 (2006) amends 10 USC 2320, "Rights in Technical Data," and requires

> program managers for major weapon systems and subsystems of major weapon systems to assess the long-term technical data needs of such systems and subsystems and establish corresponding acquisition strategies that provide for technical data rights needed to sustain such systems and subsystems over their life cycle.

The assessment is to be done before contract award and is to consider priced contract options for the future delivery of technical data.

Key DoD Directives and Instructions That Affect CLS

The preceding section identified the key legislation affecting CLS use: the requirement to identify core maintenance capabilities, limit contracting of depot maintenance to 50 percent of the workload, and award noncore work to the lowest-cost source, whether public or private. DoD has directed its components to implement these laws.

DoD Directive (DoDD) 4151.18, *Maintenance of Military Material*, promulgates the requirements of 10 USC 2464 for inherently governmental and core capability requirements and noncore capability requirements under competitive sources in accordance with 10 USC 2462 and 10 USC 2466. Maintenance program management is to begin when program acquisition activities begin, and core depot capability requirements are to be identified as early as possible in the acquisition life cycle. The directive requires establishing core capabilities no later than four years after initial operating capability (IOC). It also requires identifying core capabilities for individual components and calculating the associated depot workloads.

DoD Instruction (DoDI) 4151.20, *Depot Maintenance Core Capabilities Determination Process*, implements policy, assigns responsibilities, and prescribes procedures in accordance with DoDD 4151.18 and 10 USC 2464. This instruction specifies that the size of the core workforce accommodate the required workloads within the time limits of specified Joint Chiefs of Staff scenarios. Core capabilities and the related workloads must be able to adjust to changes in force structure, technology, or doctrine; introduction of new weapon systems; aging or modification of existing weapon systems; etc. Core requirements will be reviewed at least biennially or more often if necessary or appropriate.

Other DoD directives, instructions, and regulations affect the use, management, and reporting of CLS. Appendix B summarizes some of this additional direction.

Key Air Force Direction That Affects CLS

The Air Force has issued directives and instructions that implement congressional and DoD direction and that specify processes and organizations for implementing such direction.

Air Force Policy Directive (AFPD) 20-5, *Air Force Product Support Planning and Management*, establishes the framework for implementing product-support management in the Air Force. Product support is to begin early in the acquisition phase of a weapon system, preferably during concept and technology development, and is to provide a seamless transition to sustainment. The directive states that the primary focus of product support is on optimizing customer support and achieving maximum weapon system availability at the lowest total ownership cost. It directs program managers to document the strategy in a product-support management plan, which is considered a living document that should be updated as the weapon system progresses through its successive phases. This directive is being superseded by a new AFPD, according to comments we received in March 2008 from the Office of the Deputy Assistant Secretary of the Air Force, Acquisition Integration.

Air Force Instruction (AFI) 63-107, *Integrated Product Support Planning and Assessment*, places responsibility for both acquisition and sustainment product support planning on the program manager. This instruction addresses the life-cycle management plan (LCMP), which integrates the sustainment strategy documented with the product-support management plan and the acquisition strategy. Chapter 5 of the instruction discusses the source-of-repair assignment process (SORAP), which the Air Force uses to allocate its depot workloads, at length. The program manager initiates the process; Headquarters AFMC (HQ AFMC) provides the program manager a determination of core maintenance requirements and other assistance during the process.

AFI 63-111, *Contractor Support for Systems, Equipment and End Items*, defines the various kinds of contract support, including CLS. This instruction calls for the program manager to prepare a brochure on CLS requirements, identifying budget needs by task. It also mandates using contract support for training devices.

AFI 21-102, *Depot Maintenance Management*, assigns responsibilities for depot maintenance management to Headquarters U.S. Air Force, Logistics (HQ USAF/A4/7[3]) and HQ AFMC. HQ AFMC is

---

[3] Formerly HQ USAF/LG.

responsible for assessing minimum organic depot maintenance requirements and for determining depot maintenance sources of repair in accordance with the criteria of DoDD 4151.18. The instruction requires HQ AFMC to establish a comprehensive depot maintenance program with each ALC for all new system acquisitions, including logistics management for the life of each system.

Implications of Laws and Regulations That Affect CLS Use

Congress has long had an interest in government depots because of the jobs and economic activity they provide. Current law requires that government depots perform at least 50 percent of the Air Force's depot maintenance and that the government retain a core logistics capability able to meet both peacetime and surge maintenance requirements. These laws effectively limit how the Air Force uses CLS. The Air Force is finding ways to use CLS and still meet both core and 50-50 constraints by having government depots perform work on CLS programs so that the workload counts as organic. Such public-private partnerships are allowed under 10 USC 2474, "Centers of Industrial and Technical Excellence: Public-Private Partnerships" (see Appendix B). The Air Force's process for identifying sources of repair for new programs begins with a determination of whether laws allow room for a given program to consider contractor sources of repair. If so, the Air Force can explore these sources and choose CLS if it is a better value to the government than organic support. The Air Force violated the 50-50 law earlier this decade and has funded its depot work close to the limit since then. Barring an unlikely change in these laws, the Air Force will continue to need to choose carefully when and for what kind of work it uses CLS and, when it does use CLS for depot work, will need to be able to prove that it is the most cost effective source of repair. The DoD and the Air Force have promulgated directives, instructions, and regulations to implement congressional intent regarding CLS and to institute processes for selecting and managing CLS.

CHAPTER THREE

Funding and Cost Issues Associated with CLS Use

This chapter begins with a financial overview of contract support and CLS funding for weapon systems in the Air Force, then narrows the scope to CLS funding by type of weapon system (i.e., aircraft, missile, munitions, other, space). While the discussion will include observations about CLS costs that are applicable to Air Force programs in general, our analysis is based on an examination of the types of tasks and associated costs for the 14 largest Air Force aircraft CLS programs reported in AFTOC for FY 2006.

We will address these specific questions:

- How does the Air Force define CLS for funding and costing?
- What programs use CLS ?
- How much does each program spend on CLS, and how much in total?
- Why are CLS costs growing?
- What are the key cost drivers for CLS?[1]

Because insufficient data were available, some of these questions could not be answered to our satisfaction. This raised concerns about limitations on the data needed to manage CLS; the suggestions for CLS policy improvements in Chapter Seven reflect these concerns.

---

[1] Specifically, we explore this question in terms of element of O&S cost, such as flight line maintenance, depot-level reparables (DLRs), and aircraft or engine overhaul.

Findings and Observations: CLS Funding and Costs

How Does the Air Force Define CLS for Funding and Costing?

CLS is a subset of the more-inclusive term *contract support*, which includes various types of support for systems, equipment, and end items that contractors supply. The Air Force financial system uses several element of expense investment codes (EEICs) to refer to contract support funding. AFI 63-111 defines contract support as including pre-operational support (POS), ICS, CSS, and total contract training. The instruction explains that CSS can be provided for many logistics tasks or sustainment elements that organic logistics organizations would otherwise perform and for all or part of the weapon system. When the CSS is for multiple sustainment elements, the funding is normally coded as EEIC 578, CLS. This chapter takes a brief look at funding for CSS to provide context; the rest of the monograph focuses on CLS, defined as the efforts funded by EEIC 578.[2]

Most of the contract support costs reported in AFTOC that are identifiable as such are in the following EEICs:

- EEIC 554—Critical Space Contract Operations
- EEIC 555—Critical Space Operations–Direct Support
- EEIC 570—Contractor Operated Installations
- EEIC 578—Contractor Logistics Support
- EEIC 583—Sustaining Engineering by Contract
- EEIC 592—Miscellaneous Contract Services
- EEIC 594—Contract Technical Data.

Figure 3.1 illustrates the funding for a broad range of CSS associated with weapon systems in the Air Force over the past several years. In Figure 3.1, CLS reflects EEIC 578; Contract Services includes EEICs 570, 583, 592, and 594; and Space Operations includes EEICs 554 and 555. Figure 3.1 does not include CSS that is not associated directly with weapon systems, which support the primary mission of the Air

---

[2] To obtain the data, we queried AFTOC by weapon system and EEIC. The data reflect obligations in each fiscal year and exclude costs funded by procurement accounts for which EEICs are not available. Unless otherwise noted, AFTOC is the source of the cost information in this chapter.

Figure 3.1
Air Force CSS for Weapon Systems

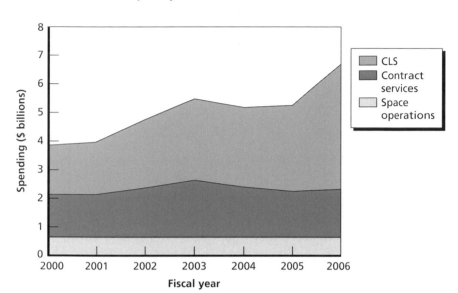

SOURCE: AFTOC OLAP, query by type of weapon system and EEIC, January 8, 2007.
NOTES: Constant FY 2004 dollars.
Data shown do not include contract sustainment support for indirect costs not
attributable to weapon systems because they have little CLS.
RAND MG779-3.1

Force. The figure does include indirect costs, those for activities that
are identifiable with an installation, site, or location, such as base operating support. Indirect costs include little CLS, although their cost of
contract services has risen significantly over the last several years.

Figure 3.1 shows that, while the funding for space operations and
contract services for weapon systems has remained fairly constant over
the past few years, funding for CLS has grown rapidly, by more than
$2.6 billion from FY 2000 to FY 2006. Space operations include con-
tractor support other than CLS. Space operations are a small subset of
CSS and are managed separately from other programs, so this monograph does not address them, except to briefly contrast that support
with aircraft CLS. Contract services for weapon systems are mostly for
aircraft systems. Their funding levels are largely stable and represent
a mix of services, including contractor operated facilities and instal-

lations, sustaining engineering by contract, engineering management support services, and program management support. We have also excluded these elements from this study.

At least two other EEICs contain contract support. One is EEIC 579, Interim Contractor Support. However, AFTOC reports almost no spending for this EEIC. Our experience suggests that, because ICS is funded through procurement appropriations, it is not recorded in AFTOC. The omission is significant because programs spend a lot of money on ICS before the transition to CLS or organic support. For example, in FY 2005, the C-17 program spent more than $900 million on ICS, and the F-22 program spent $377 million on "Performance-Based Agile Logistics Support," both funded from the Aircraft Procurement appropriation.[3] This highlights a difficulty cost estimators face when attempting to identify all sources of sustainment funding.

A second code that contains some contract support is EEIC 54*, Purchased Equipment Maintenance (the asterisk indicates that this two-digit code has many sublevels, such as EEIC 543 for purchased engine maintenance). EEIC 54* is mostly depot maintenance and is supposed to reflect purchases from within DoD and other government organizations, not contract support. In reality, government depots accept funding from Air Force customers and subcontract some of the work to contractors. The reverse also happens when contractors subcontract some of their effort to government depots. In both cases, the accounting is lost at this level of reporting; as a result, these numbers only roughly approximate whether government or contractors actually performed the work. Purchased Equipment Maintenance totaled $2.7 billion in FY 2006. It is not possible to tell from AFTOC data how much of this amount may have been subcontracted to contractors. These examples highlight the difficulty of putting together a comprehensive picture of depot-level contract support for Air Force systems.

EEIC 578 for CLS is used when CSS is intended to extend through a weapon system's entire life cycle and, generally, for multiple sustainment elements. A classic example of CLS use was for the F-117 program: The contractor provided some squadron-level maintenance,

---

[3]   These numbers are in then-year dollars and are taken from U.S. Air Force, 2006.

repair and replenishment of parts, depot maintenance, sustaining engineering, and supply chain management, among other tasks. In short, the F-117 contractor provided virtually every kind of logistics support that organic Air Force elements would otherwise provide. Other codes for contract services, primarily the EEICs listed earlier, are used for interim, rather than permanent, CLS; for selected, limited elements of contract support when most of the support is provided organically; and for CLS for space systems.

The definition of CLS is important because it excludes significant spending on similar contract support across the Air Force as a whole, as well as on significant portions of individual programs. Most significantly, this definition of CLS excludes ICS funded by the Aircraft Procurement appropriation used to support programs including the C-17 and F-22, as mentioned earlier. The large increase in CLS funding between FYs 2005 and 2006 reflects in large part the transition of support for these two programs from ICS to CLS.

More idiosyncratically, on some individual programs, such as the B-2 program, some of the CSS is coded as CLS but most is not, even though the same type of support is coded as CLS on other programs. So, in this case, the identification of CLS by EEIC is not consistent across programs. For the B-2 program, the coding of CLS changed in FY 2007, so AFTOC shows CLS as such.

The definition of CLS used in this monograph includes only work coded as EEIC 578. This narrow definition was used to enable identification of CLS costs in AFTOC and because, without intimate knowledge of how every aircraft program in the Air Force is sustained, we had no other way to identify CLS programs.

Figure 3.2 shows that, of the $4.4 billion the Air Force spent on CLS for aircraft, space, missile, munitions, and other systems in FY 2006, 86 percent was spent on aircraft systems (including engines and avionics) according to this definition.

Aircraft systems account for a large majority of total O&S costs and of CLS costs in the Air Force. While a majority of total Air Force CLS funds are spent on aircraft systems, space systems rely to a greater extent on contractor support for their operations and maintenance (O&M). This is consistent with our discussions with satellite program

Figure 3.2
Air Force CLS Spending by Type of System, FY 2006

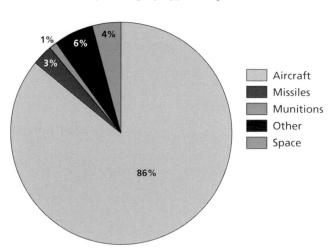

SOURCE: AFTOC OLAP, query by type of weapon system and
EEIC, January 8, 2007.
RAND *MG779-3.2*

office personnel, who told us that space systems consist of three seg-
ments: space, control, and ground. They said that the Air Force always
relies on contractor support for the space segment, almost always uses
contractor support for the control segment, and usually uses contractor
support for the ground segment. Most of the space-system contractor
support is reported in EEICs 554 and 555. More than 50 percent of
the O&S cost of space systems is attributable to contractor services. A
perusal of CLS brochures for space systems indicates that almost all of
the contractor effort is contractually fixed each year. This is consistent
with the on-off nature of space systems and the requirement for them
to be operational at all times.

Funding for aircraft systems is more closely tied to their usage (that
is, how much they are flown) than is funding for munitions, space, or
missile systems. Significant portions of aircraft O&S funding, such as
fuel, consumables, and DLRs, are determined largely by flying hours.
In contrast, munitions and missile systems are maintained and kept in
a state of readiness to be used, and space systems are either operational
or not.

Because a large majority of Air Force CLS funding is spent on aircraft systems and because their funding is more linked to actual system use and is therefore easier for Air Force decisionmakers to adjust, we will focus on aircraft CLS programs in the remainder of the monograph.

Figure 3.3 provides further insight into CLS funding for aircraft systems. It illustrates the proportions of funding and amounts spent on the largest elements of aircraft O&S costs, such as military and civilian personnel costs and DLR costs, in FY 2006. The Air Force spent $30.9 billion on O&S for aircraft systems in FY 2006. The data from AFTOC show that the largest element of O&S cost for aircraft systems is military personnel, at 30 percent of the total. Most of these are maintenance personnel. The second-largest element of cost in FY 2006 was CLS, at 12 percent of the total, followed closely by aviation fuel and organic repair and purchase of DLRs.[4]

Figure 3.3
Air Force Aircraft Operating and Support Costs, FY 2006

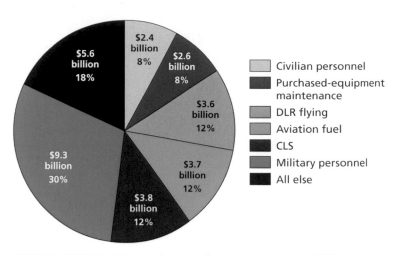

SOURCE: AFTOC OLAP query by type of weapon system and EEIC, January 8, 2007.
NOTE: Constant FY 2004 dollars.
RAND MG779-3.3

---

[4] Costs shown in Figure 3.3 and described here are from an AFTOC OLAP query by weapon system, EEIC, and fiscal year (January 8, 2007). CLS costs reflect EEIC 578 only. DLR costs include the organic cost to purchase and repair reparable items.

What Programs Use CLS, and How Much Does Each Spend on CLS?

Figure 3.4 shows the 14 largest aircraft CLS programs in FY 2006 and their growth since FY 2000.[5] The total value of these CLS contracts in FY 2006 was $3.1 billion, or 83 percent of the Air Force CLS total for aircraft. To give a more complete picture of the contractor support for the programs, that figure also includes the associated ICS. The C-130 and Global Hawk programs included relatively small amounts of ICS,

Figure 3.4
Top 14 Largest Air Force Aircraft CLS Programs in FY 2006 and Their Growth Since FY 2000

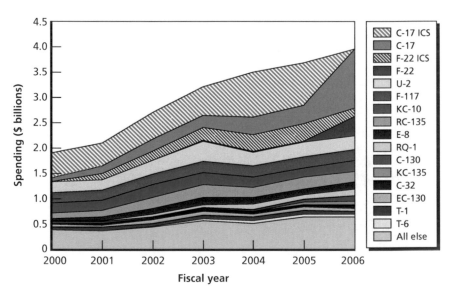

SOURCE: AFTOC OLAP query by type of weapon system, aircraft mission-design, and EEIC, January 2007, supplemented with ICS costs from Air Force aircraft procurement budgets.
NOTE: Constant FY 2004 dollars.
RAND MG779-3.4

---

[5]   The 14 largest programs were selected for illustration because they represent a broad cross-section of aircraft types and accounted for more than 80 percent of aircraft CLS spending in FY 2006. Our definition of CLS programs as those funded by EEIC 578 excludes some CLS effort. A particularly noticeable example is the B-2; while this program relies heavily on CLS, EEIC 578 funds only roughly $40 million a year of that effort. A broader definition of CLS using other EEICs would capture additional CLS costs on the B-2 program but would include many other efforts that are not CLS on other programs.

but this funding was especially important for the C-17 and F-22 programs, which relied heavily on ICS through FY 2005.

In FY 2006, the C-17 program transitioned from ICS funded by the procurement budget to CLS. The C-17 CLS contract obligated more than $1.1 billion in FY 2006 and is now by far the largest CLS program in the Air Force. Also in FY 2006, much of the F-22 program's logistics support changed from ICS to CLS, quickly making it the second-largest CLS program.

Table 3.1 shows the top 14 largest aircraft CLS programs in FY 2006 and their annual funding in FYs 2005 and 2006. The table

Table 3.1
**Fourteen Largest Air Force Aircraft CLS Programs
(FYs 2005 and 2006)**

Rank in FY 2006	2005		2006	
	Program	Funding ($M)	Program	Funding ($M)
1	C-17	360.2	C-17	1,163.8
2	F-22	15.7	F-22	380.6
3	U-2	288.3	U-2	275.2
4	F-117	214.8	F-117	217.0
5	KC-10	185.1	KC-10	210.4
6	RC-135	224.0	RC-135	201.8
7	E-8	126.1	E-8	154.1
8	RQ-1	76.3	RQ-1	125.9
9	C-130	67.7	C-130	104.3
10	KC-135	43.1	KC-135	75.8
11	C-32	42.2	C-32	61.0
12	EC-130	66.2	EC-130	58.2
13	T-1	74.9	T-1	58.1
14	T-6	51.9	T-6	56.1

SOURCE: AFTOC OLAP report for EEIC 578, data by weapon system as of January 2007.

NOTE: Constant FY 2004 dollars.

allows easier identification of the funding amounts and highlights the changes between the two years. The table also shows that the transition from ICS to CLS for the C-17 and F-22 programs accounts for much of the growth between the two years.

Programs That Account for the Growth in CLS Over the Last Seven Years

In Chapter One, we highlighted the rapid growth in CLS costs over the last seven years. CLS use has increased in all kinds of Air Force systems—space, munitions, missiles, aircraft, and other—but most of the cost growth has been in aircraft systems.

Figure 3.5 shows the rapid growth in *aircraft* CLS funding only, compared to *aircraft* organic maintenance funding, expressed in percentages. FY 2000 is the baseline of 100 percent for both CLS and organic maintenance funding, and the figure shows changes in the rel-

Figure 3.5
Growth in Aircraft CLS and Aircraft Organic Maintenance Funding

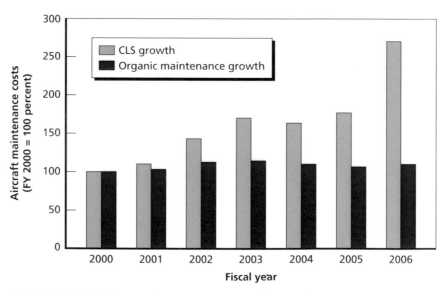

SOURCE: AFTOC OLAP query by weapon system and CAIG element, May 2007.
NOTE: Organic maintenance costs are approximated as total aircraft operation and support costs less crew, fuel, CLS, and indirect support.
RAND *MG779-3.5*

ative funding over time. Similar to the total for all Air Force systems, the organic maintenance funding of aircraft has risen slightly since FY 2000, while the funding of aircraft CLS has increased roughly 170 percent in constant FY 2004 dollars.

Figure 3.6 shows the CLS funding for the 11 individual aircraft programs that grew the most from FY 2000 to FY 2006 in constant dollars. In total, these programs account for more than $2 billion in real growth over the period, with the C-17 alone accounting for more than $1 billion in growth. ICS funding for the C-17 and F-22 programs is highlighted in cross-hatched patterns to place the growth of CLS funding for those programs in context. Figure 3.6 excludes some of the largest CLS programs, such as the F-117, because they were not

Figure 3.6
Aircraft CLS Program Costs, Largest Cost-Growth Programs

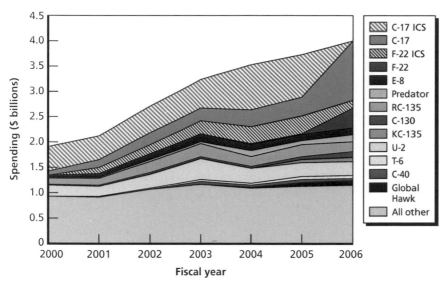

SOURCE: AFTOC OLAP query by type of weapon system, aircraft mission-design, and EEIC, January 2007, supplemented with ICS costs from Air Force aircraft procurement budgets.
NOTES: The new CLS-supported aircraft systems with increasing inventory levels above are the C-17, F-22, E-8, RQ-1 (Predator), C-130J, T-6, C-40, and RQ-4 (Global Hawk). The existing aircraft systems with no inventory growth but real growth in CLS cost are the RC-135, KC-135, and U-2.
RAND *MG779-3.6*

among those with the greatest cost growth. In the following discussion, we will assess the nature of these programs and try to determine an overall theme behind the increasing use of CLS to support aircraft systems.

Figure 3.6 ranks the programs in order of growth, those with the greatest growth are at the top. Most of the 11 CLS programs support new aircraft systems with increasing inventories; a few support existing aircraft systems that have had real growth in CLS funding.[6] While much of the overall funding growth in this period resulted from increasing aircraft inventories, the increase shown between FYs 2002 and 2003 was also affected by Operations Enduring Freedom and Iraqi Freedom.

Figure 3.7 illustrates the growth in inventory of the CLS aircraft. The force structure of both unmanned aerial vehicle (UAV) systems, Predator and Global Hawk, grew rapidly during this period. Official inventory numbers for the UAVs are difficult to obtain, so the inventories displayed reflect an approximation based upon deliveries and attrition. The Global Hawk inventory will continue to grow for several years.

Of the 11 aircraft systems with significant growth in CLS costs, three are manned surveillance-and-reconnaissance aircraft (E-8, RC-135, and U-2) with sophisticated electronics and small fleet sizes, and two more are unmanned surveillance-and-reconnaissance aircraft (Predator and Global Hawk). Four of the 11 aircraft are relatively simple commercial derivatives (C-40, KC-135, T-1, and T-6[7]). The two remaining aircraft, C-17 and C-130J, represent atypical uses of CLS.[8] The C-17 has a commercial-derivative engine, but the airframe was developed for its military mission. The CLS contract supports both the airframe and engine. While Lockheed hoped to make both commercial and military sales when it developed the C-130J, the airframe has

---

[6] Some of the growing programs are still relatively small compared to aircraft with stable funding in the "other" category, including the F-117.

[7] The T-6 has been extensively modified from the original Pilatus PC-7 trainer aircraft and has had no commercial sales. See Lorell, Graser, and Cook, 2005, for a discussion on this.

[8] See Lorell, Graser, and Cook, 2005, for case studies of C-17 and the C-130J.

Figure 3.7
Inventory of Aircraft with Growing CLS Programs

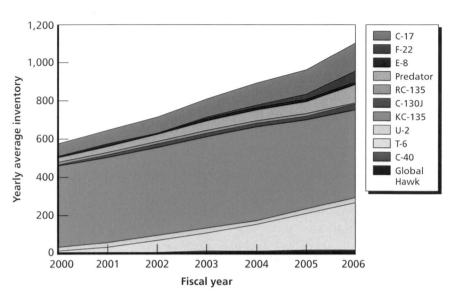

SOURCE: AFTOC OLAP, manned aircraft inventory, November 2006.
NOTE: Inventory for unmanned systems is approximated by cumulative deliveries
reported in Air Force Procurement budgets less attrition.
RAND *MG779-3.7*

many features unique to its military missions. There have thus been no
commercial sales, although there have been sales to foreign militaries.
The CLS contract supports the parts of the aircraft unique to the "J"
configuration.

What Are the Key Cost Drivers for CLS?

There is a great deal of variation in the nature of the CLS tasks for different programs. The difference in the tasks is due in part to the type
of weapon system supported: space, missile, or aircraft. But even for
aircraft systems, the nature of the support still varies a great deal from
one program to another.

Before looking at the nature of support across programs, it is
useful to review a standard way of organizing and referring to O&S
costs. OSD CAIG has defined a cost-element structure that categorizes
and defines O&S costs for DoD systems. OSD CAIG (1992, p. 4-2)

provides the basic structure and defines the elements. The following is a slightly abridged version:

1.0 Mission Personnel
    1.1 Operations
    1.2 Maintenance
    1.3 Other Mission Personnel
2.0 Unit-Level Consumption
    2.1 Petroleum, Oil, and Lubricants (POL)/
        Energy Consumption
    2.2 Consumable Material/Repair Parts
    2.3 Depot-Level Reparables
    2.4 Training Munitions/
        Expendable Stores
3.0 Intermediate Maintenance (external to unit)
4.0 Depot Maintenance
    4.1 Overhaul/Rework
5.0 Contractor Support
    5.1 Interim Contractor Support
    5.2 Contractor Logistics Support
6.0 Sustaining Support
    6.1 Support Equipment Replacement
    6.2 Modification Kit Procurement/
        Installation
    6.3 Other Recurring Investment
    6.4 Sustaining Engineering Support
    6.5 Software Maintenance Support
    6.6 Simulator Operations
7.0 Indirect Support
    7.1 Personnel Support
    7.2 Installation Support

The guide provides additional information about the seven main elements and the additional levels of detail below them.

We analyzed CLS brochures for most of the largest aircraft CLS programs, a sample comprising roughly 80 percent of CLS spending for all aircraft, to determine the key cost drivers by element of O&S

cost.[9] The analysis was imprecise because the tasks and costs in the CLS brochures did not always fit neatly into the standard O&S cost elements, so we had to make informed allocations of ambiguous tasks into the standard cost elements. We found that depot maintenance for aircraft and engines accounted for roughly 41 percent of CLS costs. Spare and repair parts accounted for another roughly 22 percent of CLS costs. Squadron- or field-level maintenance may account for as much as 14 percent of CLS costs, although ambiguous wording in the CLS brochure for one large program affected the apportionment for this task, so the amount may be smaller. Engineering support accounts for roughly 9 percent of CLS costs. The remaining 14 percent of CLS costs cover supply support or supply management, software mainte- nance, and other miscellaneous tasks. The key drivers of aircraft CLS costs, taken from a sample of large programs in FY 2006, are shown in Figure 3.8.

In contrast to aircraft systems, which use CLS for a broad variety of tasks but mostly for depot maintenance of aircraft and engines and for the repair and purchase of parts, satellite and missile systems tend to use CLS for sustaining engineering. Only two satellite or missile programs spent more than $50 million on CLS in FY 2006, and these two large programs illustrate CLS use for satellite and missile programs in general. The Minuteman missile program spent roughly $116 mil- lion on CLS in FY 2006. Of this amount, 75 percent was for sus- taining engineering. The Military Satellite Communications program spent more than $56 million on CLS in FY 2006, with a large majority of the effort described as sustainment support or engineering services. As mentioned earlier in this chapter, space programs use a considerable amount of additional contract support categorized as contract services and space operations.

Missile and satellite programs differ from aircraft systems in the *nature* of the support, in that few if any aircraft programs use CLS

---

[9] The sample included most of the large CLS programs discussed earlier, with some excep- tions. For example, the KC-135 program is not in the sample because we did not find a CLS brochure for it, and the C-130J is not included because the contract was executed in FY 2006 at a far lower level than shown in the brochure, and we did not have insight into which tasks changed.

Figure 3.8
Key Drivers of Aircraft CLS by Element of O&S Cost

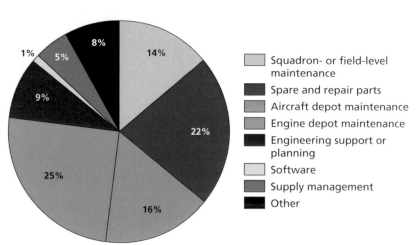

SOURCE: FY 2006 data from FY 2008 CLS brochures.
RAND *MG779-3.8*

primarily for sustaining engineering, although sustaining engineering is frequently part of the total support package. Missile and satellite programs also differ from aircraft systems in the variability of the *scope* of the CLS across O&S elements. For some aircraft programs, such as the U-2, contractors provide practically all the support, from flight-line maintenance to spare and repair parts to depot maintenance. On other programs, CLS may be used for a small fraction of total O&S costs to support one or two subsystems. For example, the F-15 program relies on CLS primarily to support the APG-63 fire control radar.

In spite of these variations, it is possible to generalize broadly about how Air Force weapon systems use CLS yet keep in mind that individual programs may differ markedly from the average. For space and missile systems, as noted earlier, CLS tends to be used to provide sustaining engineering support, rather than maintenance at the organization or depot level. For aircraft systems, most of the CLS funding is for depot overhaul and repair.

Having calculated and provided broad generalizations about what tasks CLS performs, we acknowledge that precise measurement of CLS effort by task or O&S cost element is problematic for at least two major reasons. First, the picture is changing over time, especially as the C-17 and F-22 transition from ICS to CLS. These two programs have become by far the two largest CLS programs in the latter half of this decade, so the actual costs for aircraft CLS programs will grow after FY 2006 and may look different within two or three years. The increasing use of public-private partnership arrangements for these programs and others, in which organic depots perform work that contractors plan and supervise, further blurs the distinction among tasks. A second major problem in measuring CLS costs precisely is that the CLS brochures on which we have relied to provide insight into CLS costs do not usually break out the costs by O&S CAIG element. For the C-17, for example, the description of tasks for airframe labor and materials mentions both organizational and depot-level effort but does not break out funding by each level. The proportions summarized in the preceding paragraph reflect an educated categorization of the funding amounts from CLS brochures, but some error in the categorization is inevitable.

Observations and Conclusions About CLS Funding and Costs

As a broad overview, a large majority of CLS funding is for aircraft systems, and CLS funding is growing faster than other O&S funding largely because inventories of contractor-supported aircraft are growing. Aircraft CLS is a logical place for Air Force decisionmakers looking for possible budget cuts because this is where the money is and because O&S funding is more closely related to rates of aircraft usage than it is for missiles, munitions, or space systems.

More critically, the Air Force does not collect detailed or uniform data on CLS costs. AFTOC, the official Air Force O&S management information system, contains only total CLS costs per program and offers no further detail. For insight into CLS tasks and costs, we had to rely on CLS brochures, which are budget-requirement documents that provide estimated costs for the fiscal year in which they are prepared and budget requests for future years. One of the implications for cost analysts of the data limitations is that the CLS costs will not be recorded in AFTOC in the appropriate O&S CAIG cost element, and the detailed commodity transaction costs for consumables and DLRs will not be recorded because these transactions are conducted outside of the government supply system. To have a clear picture of the costs by O&S element and/or for all equipment on the weapon system, cost analysts must be aware of the tasks CLS performs. CLS will affect the amount and proportion of costs by O&S CAIG element, and the costs displayed in AFTOC may not accurately represent the costs by element.

Our recommendations, discussed in Chapter Seven, address the shortfalls described earlier. We suggest that the Air Force require CLS contractors to report cost data to at least level 2 of the CAIG O&S Work Breakdown Structure. This will increase the availability of CLS cost data and categorize the data according to a useful standard.

CHAPTER FOUR

Assessing the Performance of CLS

In this chapter, we address issues regarding the performance of CLS in terms of cost and effectiveness:

- How does the government measure CLS performance and using what metrics?
- What has contractor cost performance been like for major programs using CLS?
- How effective has CLS proven for major programs using CLS?
- How does CLS price growth compare to organic cost growth?
- Do contractors have inherent advantages or disadvantages in performing some tasks?

We were able to answer some of these questions in more depth than others depending on the data available.

Cost and Performance: Comparing CLS with Organic Support

How Does the Government Measure CLS Performance and Using What Metrics?

We discussed the metrics program offices use to assess CLS performance with 11 of the program offices that have the largest Air Force CLS contracts. A recurring theme was that the government should measure only what the contractor can control and that it must be careful to measure and reward behaviors it wants to encourage. Because

CLS contracts for different programs buy different services, the metrics will vary accordingly. When the contractor provides total logistics support and is therefore responsible for total system performance, broad metrics of weapon system availability, such as mission capable rate, are used. However, most weapon systems do not rely completely on CLS but instead use some mixture of organic and contractor support, and it is here that it is important to select metrics that measure only what is under the control of the contractor.

Table 4.1 lists common metrics for tasks or elements of O&S costs that are typical of CLS contracts. Not all CLS contracts in any category use all the relevant metrics. Some of these metrics have common definitions:

- **partially mission capable supply**—the percentage of time an aircraft can fly at least one but not all of its missions for reasons attributed to supply
- **not mission capable supply**—the percentage of time an aircraft is grounded and cannot fly any of its missions for reasons attributed to supply
- **issue effectiveness**—the percentage of customer requests that have been filled by items in the inventory; includes the fulfillment of any request, not just requests for items the supply is authorized to stock (Air Force Logistics Management Agency, 2001, p. 44)
- **mission incapable awaiting parts**—the percentage of time that an aircraft is unable to perform its assigned mission because of a lack of parts
- **stockage effectiveness**—the percentage of customer requests filled by items that the supply system is authorized to stock (Air Force Logistics Management Agency, 2001, p. 44)
- **mean time between repairs**—flying hours divided by repair actions
- **mean time between failure**—a basic measure of reliability for reparable items; the average amount of time that all parts of an item perform within their specified limits
- **repair turnaround**—a measure of the length of time to repair an item and return it to the stock system

Table 4.1
Metrics Used to Assess CLS Performance

	CLS Service Purchased	Metrics
Supply system	Contractor operated and maintained base supply	Partially mission capable supply
		Not mission capable supply
		Issue effectiveness
		Mission incapable waiting parts
		Stockage effectiveness
	Spare and repair parts	Mean time between repairs
		Mean time between failure
		Repair turnaround time
		Break rates
		Mean time to repair
		Cost control
Overhaul	Programmed depot maintenance (PDM)	Cost
		Schedule
		On-time delivery
		Depot deficiencies
	Support equipment	Availability of support equipment and vehicles
	Total weapon system	Aircraft availability
		Flying hours achievable
		Mission-capable rate
		Subjective customer judgment
		Mission success rate
		Satisfaction of the operational commander with the contractor's performance
		Cost per flying hour
		Cost per aircraft per year

- **break rates**—the number of breaks, defined as landings with write-ups requiring major maintenance that ground the aircraft, per sortie
- **mean time to repair**—a basic measure of maintainability: the total maintenance time divided by total number of failures

- **aircraft availability**—mission-capable hours divided by total possessed hours
- **mission-capable rate**—the percentage of all possessed aircraft capable of fulfilling at least one of their assigned missions.

What Has Contractor Cost Performance Been Like for Major Programs Using CLS?

We tried to examine contractor cost performance across and within programs. We were not able to address this question well because of insufficient data. The two cost metrics identified in Table 4.1 to assess performance within programs refer to costs to repair specific parts or programmed depot maintenance (PDM) work packages, and this detailed information was not available to us with one exception noted later.

At the level of the overall weapon system, typical performance metrics include aircraft availability, achievable flying hours, and mission-capable rate (see Table 4.1). It would be possible to work from the standards for these to assess the CLS costs for achieving these, but the endeavor is not simple. Because such assessments have been and will probably continue to be of interest to those involved with logistics support of weapon systems, we provide the following discussion of the issues and difficulties in assessing CLS cost performance.

Cost performance can be judged or measured in different ways. For O&S, costs are commonly measured per unit (e.g., O&S cost per aircraft) because costs are affected by the number of units supported. Logistics costs are also often measured per unit per operating hour (e.g., aircraft O&S cost per flying hour) because time in use affects logistics support costs. Accepting and using these common measures of cost on a per-unit or per-hour basis allows assessments of cost to be considered across programs or within a program over time.

One difficulty in assessing the costs of logistics support providers across programs is that differences between weapon systems affect their O&S costs. For example, Air Force aircraft differ widely in their missions, sizes or weights, equipment, age, reliability, and use, and these differences affect their cost per unit or per flying hour, regardless of support concept. It is very difficult to normalize for all the various characteristics of weapon systems that affect costs to allow assessment of a cost difference attributable only to CLS.

Another difficulty in assessing CLS costs across programs is that, even after normalizing for inherent differences between weapon systems, the scope of support on CLS contracts varies tremendously across programs. Any comparison of CLS performance across programs would have to ensure inclusion of only relevant tasks. Such comparisons require knowledge of the CLS tasks performed and their associated costs. AFTOC does not report detailed CLS costs as it does for organic costs. CLS brochures report costs by task, but the available brochures typically cover only a limited number of years worth of data, and the descriptions of the tasks vary across brochures.[1]

In theory, the difficulties discussed earlier could be addressed using some form of statistical modeling. For example, multiple regression would hold essentially constant with respect to each other all the characteristics that affect cost to allow measurement of the cost difference attributable to each characteristic, including CLS. However, this approach is difficult in practice because the analysis would have to address a comparable set of tasks and costs, include many observations of data to normalize for all of the characteristics of programs that affect O&S costs, and quantify the characteristics accurately and adequately. These data do not exist.

Another way to assess CLS cost performance is to examine data on the same program over time. Some Air Force programs have switched from organic to contractor support, and the cost performance of each type of support for a single program could be compared. The difficulty with this approach is that some other characteristics, such as age, usage, and fleet size, affect the O&S costs of an individual program over time. Fair assessment of the performance of different support concepts for a single program must normalize for these differences. Fleet size is probably the most difficult characteristic to normalize, in part because CLS costs may include an unknown combination of fixed and

---

[1] The brochures are also required to be produced using a unique computer program, which means the data cannot be easily read and analyzed using more commonly available software.

variable costs that distorts cost per unit or per flying hour as fleet size changes. Furthermore, direct comparisons of CLS and organic costs are complicated by the tendency for organic costs to vary more depending on available funding, while CLS costs are generally more stable. In addition, many organic costs (such as overhead costs at ALCs) are distributed across several systems, so costs have to be allocated to approximate the cost of each system. In contrast, all CLS costs must be paid through the contract mechanism.

Another complication in assessing cost performance across or within programs is that cost per unit or per flying hour is not the only metric of interest to the Air Force or necessarily the most important metric in measuring logistics support. Flying hours represent time that the aircraft is used for peacetime training or in combat, and this measure of actual use is surely important. But flying hours are limited in part because they are costly. The availability of the aircraft to fly when necessary is also valuable. So, such other metrics as the percentage of time that the aircraft is capable of performing its mission, known as the mission-capable rate, are important in assessing logistics support.

Aware of these conceptual and practical difficulties involved with assessing costs, we nevertheless tried to collect relevant data and found that scant data are available, and those that are available are not centrally collected. We asked how the cost and performance for CLS compared to those for organic support, including initial assessments at the time of the source-of-repair decision or those based on performance to date. We found that no particular office was responsible for conducting business-case analyses or similar studies that would address such issues. When such analyses were done at all, individual program offices appear to have done them. The program office representatives we spoke with offered little in the way of formal studies or quantitative evidence that evaluated CLS performance relative to organic support. For most programs, it is difficult to make meaningful comparisons, and there is little if any reason to do so once a source-of-repair decision has been made. Two instances in which such comparisons are practical and make sense are when a program changes its source of repair and when a program has an ongoing choice of sources of repair for the same or similar work.

In spite of these difficulties, we were able to obtain two studies that compared organic and CLS costs. One analysis was for the KC-10 source-of-repair decision in the early 1980s. CLS was found to be less expensive for the small fleet of KC-10s, mainly because of the large nonrecurring cost of standing up an organic capability, given that the contractor had already invested in a support infrastructure for the commercial aircraft. Furthermore, since the KC-10 was derived from a true commercial aircraft (rather than being a military aircraft planned for the world defense market), there were even multiple companies that could undertake the sustainment.

The second cost comparison, or more accurately, set of comparisons, contrasts CLS, organic, and public-private depot partnership costs for various depot-repair workloads on the F-22 program. These studies, called depot partnering assessments, were ongoing at the time of this writing. They separate the depot repair work on an aircraft by workload or commodity and examine each workload to determine where it can be performed most economically. Each assessment includes an estimate of the nonrecurring cost of establishing depot repair capability under each of the three alternatives. Similar to the findings of the KC-10 study, the nonrecurring costs for tooling; test equipment; and in some instances, technical data can be a discriminator among repair alternatives. The assessments also contain a projection of recurring costs. We have not provided more-detailed descriptions of the assessments and their findings because they contain proprietary information and because program sustainment decisions were in process as of this writing.

One case did allow us to compare costs for the same work on the same program at the same time, thus avoiding the conceptual difficulties involved in making comparisons we discussed earlier. We obtained average costs per PDM for the KC-135 by contract and organic depots from FY 1994 to 2003. The cost per aircraft at both contract and organic depots roughly tripled during the period, and the price grew unevenly from year to year. In addition to cost information, the program office monitors measures of the quality of performance, such as schedule and deficiencies. KC-135 PDM is an unusual case in that the large fleet of commercial-derivative aircraft provides enough workload to allow multiple sources of repair and ongoing competition for the work.[2]

We also discussed the cost of organic and CLS with people knowledgeable about logistics support.[3] The most striking finding was the diversity of opinion. Some respondents, generally those from programs that used CLS, believed that CLS was superior in cost and performance.[4] Other respondents believed that organic support was usually less expensive. Still other respondents argued that the specific business arrangement between the program office and the repair organization, whether public or private, was the key to successful outcomes. A sampling of representative comments follows from each of the three viewpoints.

Respondents who believed that CLS was generally superior in cost and performance offered the example of PDM work on variants of the C-135, such as the KC-135 and RC-135. Both contractor and government depots perform airframe PDM on C-135 variants. Those familiar with the cost and performance of both depots claimed that the hourly "wrap" rate (the rate including labor, material, and overhead costs) is much lower at contractor facilities. This contractor cost advantage is not, however, borne out by the data on cost per PDM that we obtained and described earlier.

Other respondents were equally sure that organic support was more cost-effective than contractor support was. One person had done a comparison of depot work several years before and had found that government depots were always less expensive than the depots of

---

[2] The KC-135 program office provided the information on contractor KC-135 PDM costs in a briefing. We describe but do not show these data to protect the sensitive cost information.

[3] We talked with people in a variety of positions in the government. Many worked in program offices that used CLS, and some worked at Air Force operational commands or headquarters organizations. We also talked with people in OSD and GAO. The functional areas of the respondents included finance, logistics, contracting, and program management.

[4] It should be noted that the focus of the study meant that we spoke directly with programs that used CLS, not programs that were primarily organic. We do not know the opinions of representatives of organic programs. However, we believe their perspective was captured by the somewhat more balanced information provided by headquarters staff outside of specific program offices.

prime contractors and that private third-party depots were sometimes cheaper than organic and sometimes not. Another claimed that a fair comparison of depot work will always show a cost advantage for the government because it does not add profits or fees to its cost, which contractors do.

Still other respondents held more-nuanced positions. One claimed that the Air Force has not had the analytic capability for many years to do cost comparisons of contractor to organic support and that no credible studies have been done, so there was no way to know. A particularly difficult problem in the analysis of costs of contractor and organic depots is capturing overhead (indirect labor and material costs) to ensure a fair comparison.

Another argued that the key to successful outcomes in logistics support is setting the right terms in the business arrangement. In the case of CLS, the program office has to write the contract to a specification that forces the program office to define its requirements and stick to them. The process of contracting with industry imposes this discipline, and choosing the correct metrics is crucial. This person argued that arranging for organic logistics support does not involve the same process of defining requirements, so requirements and cost tend to increase. Advocates of performance-based logistics (PBL) might point out that the same principles of defining requirements and measuring performance could be applied to either public or private logistics support.

How Effective Has CLS Proven for Major Programs Using CLS?

We tried to obtain quantitative information that would allow us to compare the performance of CLS with that of organic support. We thought the fairest comparison would be to assess metrics for only the elements of support CLS provides against comparable organic support elements. This reasoning excludes metrics that reflect the performance of the entire logistics system, such as the operational availability of the weapon system. Instead, we tried to compare only areas for which a CLS or organic organization was completely responsible for the support and for which there were relevant metrics of performance. With these conditions in mind, we tried to obtain information on logistics performance in depot maintenance and supply systems.

Depot maintenance performance can be assessed using such metrics as cost, on-time performance, and deficiencies per overhaul. Although this information is collected and is available at HQ USAF, we were not able to obtain it in the course of our research. Judging from the variety of information emerging from our interviews and the lack of quantitative evidence we received about depot performance, few people working with CLS appear to have access to the information.

Supply-system performance can be assessed by such metrics as "not mission capable supply" rate, "mission impaired capability awaiting parts" rate, and similar measures. These measures are not perfect because unclear situations often make it possible for the data to be entered incorrectly or to be affected by the inconsistent judgment of maintainers. For example, equipment could be coded as "not mission capable supply" in one instance but as "not mission capable maintenance" in another. Although imperfect, these metrics are available from an HQ USAF/A4/7 database.[5]

A more-serious difficulty is that these metrics of supply-system performance are affected not only by the efficiency of the organizations that manage the supply systems but also by such other factors as the amount of funding available for the spare parts the organization manages. All else being equal, higher levels of funding for spare parts will result in better "not mission capable supply" rates and other measures of performance. Because it is difficult to determine the amount of funding provided for spares over the life of a weapon system, especially relative to the weapon system's need for spares, it is difficult to assess the supply-system performance fairly.

In addition, it is difficult to identify comparable costs of government and contractor supply systems. The government and contractors account for supply-system costs differently, and thorough knowledge of

---

[5]   The Multi-Echelon Resource and Logistics Information Network (MERLIN) was available online subject to restricted access until it was decommissioned in April 2008. Its replacement, Logistics Installations and Mission Support (LIMS), is available through the Air Force portal.

how each organization accounts for its costs is needed to ensure a fair comparison. The difficulty in identifying comparable costs of CLS and organic supply systems makes a cost-effectiveness comparison of their supply support problematic.

One comparison of organic and contractor depot and supply-system performance avoided the difficulties described earlier. The F-117 program switched from organic support to CLS during the period AFTOC and MERLIN covered, making funding and performance data available under each kind of logistics support. The analysis showed improvements in such metrics as "not mission capable supply," "mission impaired capability awaiting parts," and "depot delivery schedule" after the change to CLS. We cannot say for certain why performance improved after the change to CLS. It may be that, all things being equal, the contractor performed better, or it may be that more-stable funding allowed for better performance.

Despite the difficulties in identifying comparable costs of supply systems and comparing performance, we have made a limited comparison of supply-system effectiveness using the "total not mission capable supply" (TNMCS) metric; see the related discussion in Appendix A.

Discussion on CLS Supply Support

The comparisons of organic support and CLS in Appendix A indicate more-demanding TNMCS standards for CLS aircraft in every case. The cargo and tanker CLS programs achieved better absolute and relative TNMCS performance compared to organic support, while the fighter and trainer comparisons were more ambiguous. We discussed the findings with several people familiar with logistics support to gain insight into reasons for differences in logistics performance as measured by this metric. These people offered some possible explanations.

One observation is that CLS has been used to support most of the newest aircraft in the Air Force inventory, while most of the older aircraft are organically supported. All else being equal, newer aircraft tend to be more reliable than older aircraft and therefore tend to require less maintenance and fewer parts and put less demand on the supply system, which could affect the TNMCS rates. In our comparisons, this was generally, but not always, true—among the C-130s and F-16s, the older series had better TNMCS rates, and the newer trainers had break rates roughly similar to their older counterparts. But in general, aircraft age and reliability should be taken into account when comparing the performance of different supply systems.

Another insight is that well-maintained aircraft with high availability rates are largely a function of the resources devoted to maintaining them. One major reason it is difficult, if not impossible, to find truly comparable CLS and organic systems is that their levels of O&S funding tend to differ. Many CLS programs have a high proportion of contractually fixed costs that must be funded each year if the Air Force is not to violate the terms of the contracts. Because of these contractually fixed funding levels, CLS programs have been less affected by funding instability than have organically supported programs, which must often reduce funding for spare parts when budgets are cut. Higher, more-stable funding therefore allows CLS to provide better supply and other maintenance support.

One explanation advanced to explain supply support performance, as well as that of other logistics support, is that the formal contracting process involved in writing and negotiating a CLS contract allows the program office to motivate behavior it values. Many of the program office personnel that used CLS mentioned using metrics of supply-system performance, including TNMCS, to assess and reward their contractors. The personnel observed that the organic supply system is not as readily motivated.

Similar contractual incentives can be created for other logistics functions. Proponents of CLS for depot work also said that the contractor can be motivated by the terms of the contract to deliver over-hauled aircraft on time with fewer deficiencies. Fixed-price contracts for PDMs or incentive fees on CLS contracts allow the program office to create incentives for cost and performance, which the program office cannot use as easily to motivate good performance from government depots.[6]

---

[6] Program offices could reassign work from depots if the government held full rights to the technical repair data so that private contractors could use the data and if allowed by core

A fourth reason for better aircraft reliability and availability and supply-system performance under CLS is related to contractual incentives for performance but has an additional aspect. Unlike government organizations, contractors are not limited by restrictions on what they can do with O&M funding. So a contractor funded with O&M money can procure and install a new and more-reliable part, for example, that results in fewer failures and demands on the supply system, less maintenance, and better aircraft reliability. Air Force organizations in the same circumstances would require approval and a separate EEIC at each step of the process. This is an inherent advantage to using CLS that occurs when the contractor is properly motivated by the contract to achieve logistics performance goals.

How Does CLS Price Growth Compare to Organic Cost Growth?

This question differs from the previous cost comparison in that, here, we are not interested in comparing costs for a given task or function but in comparing rates of growth over time. We were not able to answer this question satisfactorily, primarily because of the limited cost history for CLS programs. For most programs, we have the CLS brochures submitted for FY 2006, 2007, and 2008—three years of cost history, which is not enough to establish trends confidently.

A less-satisfactory and less-detailed approach would be to rely on total CLS costs per program, as reported in AFTOC, and to measure the growth per program against that of comparable organically supported programs. One danger with such a comparison is that it offers no insight into the tasks performed, which may change over time. It also does not take into account the mission environment, which may cause some aircraft to be flown more intensively than others.

Given the difficulties explained earlier, we conducted the best comparison we could within the limitations of the data. We selected programs that had been using CLS long enough for cost trends to appear and that had fairly stable inventories and usage rates. The last was important because these rates can distort costs per aircraft or per

---

and 50-50 requirements. The Air Force has been close to the 50-50 limit for the last several years.

flying hour. We compared the CLS programs with organically supported programs having the same mission, stable inventories, and similar usage rates. We compared the total direct O&S funding of the programs, excluding indirect support, which grew disproportionately on
some programs.[7] The fairest comparisons we could make were between
the F-117 program and the F-15C/D, F-15E, and F-16C/D programs;
between the KC-10 and the KC-135R/T program; and, secondarily,
between the KC-10 and the C-130 and C-5 programs.

We analyzed constant FY 2004 dollars from AFTOC from FYs
1996 to 2005. We included direct O&S costs (CAIG elements 1–6,
excluding indirect support) and calculated direct O&S costs per flying
hour to account somewhat for changes in flying hours.

The fighter programs seemed to offer the best comparison,
although the F-117 is a stealth aircraft and the others are not. The fleets
compared have roughly similar average ages and usage rates over time.
The F-117 program had by far the lowest cost growth per flying hour
of the fighter fleets. In fact, the F-117 contract was designed to create
incentives for the contractor to meet logistics performance metrics at
a constant price over time. Its cost per flying hour, in constant dollars,
grew in the single digits over 10 years, while the cost per flying hour of
the organically supported fleets grew variously from over 20 percent to
over 40 percent in real terms.

The tanker programs offer a poorer comparison because their
usage patterns are not as stable and comparable and because the KC-135
program uses contract depot maintenance for a significant part of its
rapidly growing PDM program. Since FY 2002, because of increased
wartime operations, flying hours per KC-135 have risen quickly, while
the KC-10 has not sustained as much of an increase. Total fleet flying
hours for the KC-135R/T increased 46 percent from FYs 1996 to 2005,
and flying hours per aircraft increased 24 percent during the period.
KC-10 total fleet flying hours increased only 12 percent from FYs 1996
to 2005. In general, higher usage rates tend to reduce costs per flying

---

[7]   Indirect costs are those not directly attributable to a weapon system but that are identifiable with an installation, site, or location. Base operating support is an example of indirect
costs. Indirect costs averaged seven percent of total O&S costs on these programs.

hour, and this was advantageous to the KC-135 in the latter part of the period.

KC-135R/T cost per flying hour increased 14 percent in constant dollars while the KC-10 cost per flying hour increased roughly one-quarter over the period. KC-135R/T costs increased significantly more than KC-10 costs did in absolute terms over the period, but KC-135 fleet flying hours grew more also. The tanker comparison is ambiguous, and the conclusions to be drawn depend on the interpretation of the data.

A secondary comparison can be made between the KC-10 and the organically supported C-130 and C-5 programs. The latter are cargo, rather than tanker, aircraft but, like the KC-10, are large and relatively simple platforms that offer a reasonable comparison. The KC-10 had much less cost growth per flying hour than did the C-130 program and more than did the C-5 program.

In summary, few good comparisons of price growth can be made with weapon system–level data of similar types aircraft over time. The fighter comparison shows excellent cost performance on the F-117 CLS contract, while the tanker comparison is ambiguous.

Do Contractors Have Inherent Advantages or Disadvantages in Performing Some Tasks?

In a previous subsection, we addressed supply-system performance, and observed that contractors have an inherent advantage in performing work using O&M funding without the restrictions that limit government organizations. In this subsection, we examine whether there are other tasks or types of logistics support in which either the government or contractors have an inherent advantage. This assessment is based on discussions with Air Force personnel and CLS contractors.

One advantage of using CLS for supply support is that it does not suffer the restrictions the Air Force has imposed on organic organizations by managing funds using EEICs, which limit the ways programs can use the funds from a given appropriation. Contractors that also support commercial systems also have advantages when providing supply support for military systems derived from the same systems. A CLS contractor for a true commercial-derived military system can use its inventories of spare parts for both its commercial and military aircraft, thereby obtaining better prices and taking advantage of the existing supply system. The arrangement may also offer opportunities for improvement; for example, the contractor may encounter and resolve part reliability problems in the commercial workload that would provide valuable insights for the military system.

The advantages and disadvantages are mixed for squadron- or organization-level maintenance. Several aircraft CLS programs use field support representatives (FSRs), who are contractor maintenance personnel. Their use varies across programs. In some cases, a handful of FSRs train Air Force maintainers, who comprise most of the maintenance workforce, or provide troubleshooting. At the other extreme, such as on the U-2 program, FSRs are used extensively to provide organizational-level maintenance.

One advantage of FSRs that we heard consistently in our discussions was that they are highly experienced on the weapon system they support. Their experience can be particularly advantageous on complex or specialized systems. In contrast, the Air Force is constantly developing its workforce by providing on-the-job training to its junior maintainers. As a result, at any given time, a large percentage of the military maintenance workforce is fairly inexperienced, and the more-experienced personnel are often training rather than performing maintenance tasks. In addition, military maintainers typically work on more than one aircraft model during their careers and thus do not develop the level of expertise FSRs do with a single model. Military personnel also have other duties to perform in addition to aircraft maintenance.

Using military personnel for organization-level maintenance may be advantageous when the squadron deploys. Deployment to harsh or dangerous environments is part of the military job description. Enticing contractors to deploy may require costly incentives.

We did not find a clear inherent advantage or disadvantage for either a contractor or organic source of repair for depot maintenance. A stable Air Force workforce with deep expertise performs depot maintenance. As mentioned previously, we collected a wide variety of information on the cost and quality of the work of both kinds of organizations during our interviews. We could identify only two inherent contractor advantages: One was potentially indirect and was associated with contractor supply systems; the other was the ability to use a contract to structure incentives for performance. The more-flexible contractor supply systems could provide an advantage by supplying needed parts during overhauls, which could help reduce flow time through depots. Metrics of depot maintenance performance, such as flow time or deficiencies reported, can be measured and rewarded by incentives in the CLS contract. Again, without access to metrics on depot performance, it is impossible to assess these potential advantages.

Two additional interrelated areas of logistics support that are increasingly important as weapon systems age and as the pace of technological change accelerates are obsolescence management and configuration management. It is sometimes difficult to find spare or replacement parts for older weapon systems or to buy them at reasonable prices. Actively addressing the problem by developing new sources of supply or modernizing the equipment is known as *obsolescence management*. A related issue that often arises as the same basic platform is kept in service with changing subsystems and capabilities is *configuration management*. A CLS contractor, particularly the OEM, has inherent advantages in both areas. With fewer restrictions on what it can do with O&M funding than a government organization, a contractor has greater flexibility in addressing such issues as whether to buy spares or replace old components. In addition, an OEM that provides CLS already has to manage the configuration of the items it is producing and must manage its suppliers. It is already performing these functions as a producer and may therefore also be better able to perform them as a logistics provider than an organic organization.

Observations on CLS Performance

Air Force program offices use a variety of common logistics metrics to measure and reward the performance of their CLS contractors. Program offices try to use metrics that capture only what is within the control of the CLS contractors on a given program in an effort to measure and reward the behavior they want to encourage.

We tried to assess the performance of CLS on programs over time and in comparison to other programs and found it difficult to do so. It appears that program offices seldom perform and/or retain analyses of the cost-effectiveness of the logistics support for their programs. It is difficult to make comparisons across programs because the Air Force does not collect uniform and detailed cost data on CLS; performance data for government depots is difficult to obtain; and the Air Force does not appear to collect performance data for contractor depots centrally. Furthermore, making comparisons across programs is conceptually difficult because they are affected by characteristics unrelated to the logistics support provider, such as the inherent reliability of the weapon system, its age, usage rate, and level of funding.

The analyses, information from interviews, and objective measures for comparing the costs of CLS and organic support gave mixed results. Our comparison of supply-system data revealed that CLS programs are performing well in comparison with organic systems, with the caveats that supply support is a function of funding and that the funding for the CLS programs may have been more stable or more generous than that for their organically supported counterparts.

Conclusion

Two of our findings have potential implications for the management of logistics support in the Air Force. First, experienced program personnel argued that key to successful outcomes in logistics support was the process of defining requirements, writing a contract to them, and providing financial and contractual incentives to reward performance. This process is required for CLS contracting; the same process would be useful when arranging for organic logistics support, although not necessarily to the same extent.

Second, contractors are not as limited by restrictions on what they can do with O&M funding as government organizations are. A government source of repair in the same circumstances is hampered by the restrictions implied by EEICs and would require the approval and involvement of different organizations to shift funding during a fiscal year. This inherent advantage of using CLS is a result of properly motivating the contractor through the contract to achieve logistics performance goals. This flexibility is an advantage in managing obsolescence. It would be desirable for government organizations to have similar flexibility.

CHAPTER FIVE

CLS Management

This chapter examines the Air Force's process for managing CLS, from the initial planning for the logistics support of a new weapon system to the ongoing management of a CLS contract. We address the following key issues:

- What are the current processes for choosing CLS and ongoing CLS management?
- Why was CLS chosen for existing programs?
- How are existing CLS tasks defined and funded?
- Why do CLS contracts have so little variable funding?
- How much insight does the government have into the contractor's costs?
- How are CLS contract prices determined?

Current Processes for Choosing CLS and Ongoing CLS Management

The official Air Force process for planning product support, including the process for choosing sources of repair, is described in AFI 63-107 (2004). Chapter 7 of AFI 63-107 summarizes the roles and responsibilities of Air Force organizations in planning for product support. These describe a decentralized chain of responsibility in which the program manager has primary responsibility for most product-support planning functions. The program manager is responsible for the program performance and the overall health of the weapon system he or she manages but is also responsible for meeting broader Air Force objectives, including ensuring "that the individual system or product-support strategy process is linked to all other related product support strategies to ensure support strategies are synchronized" (AFI 63-107, 2004, p. 24).

The responsibilities of the organizations at the top of the Air Force hierarchy are fewer and less clear. The instruction states that the Office of the Assistant Secretary of the Air Force, Acquisition (SAF/AQ) issues policy, with the only responsibility in planning or implementation being "Reviews LCMPs for acquisition-related product support planning" (AFI 63-107, 2004, p. 20).[1] Similarly, the roles and responsibilities of the Deputy Chief of Staff for Installations and Logistics (HQ USAF/A4/7[2]) are more numerous but consist generally of issuing policy, reviewing, and advocating during the programming and budgeting process.

The instruction calls for the program manager to develop an LCMP. The program manager is given primary responsibility for many tasks, including initiating and completing the SORAP and reporting depot maintenance obligations. AFMC is responsible for supporting the program manager in the planning and implementation of the LCMP and for the review and concurrence or nonconcurrence on SORAP recommendations. In practice, the staff in AFMC/A4B, the specific AFMC office that performs these duties, says that it helps the program manager develop the SORAP. The Acquisition Strategy Panel (ASP), when convened, approves the SORAP recommendation.

AFI 63-111 (2005) provides additional instruction on using contractor support. This instruction describes the various types of contract support, of which CLS is one. It identifies the system program manager, in collaboration with other organizations, as being responsible for identifying contract support requirements. It states that contract support "shall, to the maximum extent, be consistent with AF format standards and be compatible with AF management and data collection

---

[1] The LCMP is a combination of the former single acquisition management plan and the product-support management plan into one document that covers the life cycle of the weapon system.

[2] Formerly HQ USAF/IL.

systems" (AFI 63-111, 2005, p. 3). It also states that contract support for the life of the system is mandatory for all training devices unless HQ USAF/A4/7 has approved a waiver.

Reasons Existing Programs Use CLS

We examined the history of CLS programs for lessons applicable to future efforts. One question we examined was why CLS was selected for particular programs. A few common reasons were given.

Aircraft originally designed for *commercial* service but used in the Air Force are often supported by CLS. Three groups of aircraft are labeled "commercial." The first includes what are referred to as *near derivatives* of aircraft that commercial firms use, such as the KC-10, which is based on the DC-10. Commercial contractors are available to support these aircraft. The second group includes what are called *commercial derivatives*, even if the modifications are so extensive that they require a new sustainment approach. Here, the reasonableness of CLS is less clear, but it is typically used anyway. The third group consists of aircraft deemed *commercial* because they were intended to be sold to a broad market, that is, to more customers than just the U.S. military. For example, Lockheed Martin, which paid for the development of the C-130J, hoped to sell it to non-U.S. military customers. CLS is used for aspects of the aircraft that are unique to the J program.

Fleet size is another common consideration. Managing sustainment requires some significant investments—tooling, purchase of technical data, a supply system, and so forth. Given that the OEM has to make these investments for production and/or ICS, performing the work organically would require duplicating the OEM's investments. For a small fleet, these investments would require a larger percentage of the sustainment costs than they would for a large fleet, so this duplication is a proportionally greater penalty for a small fleet than for a large fleet.

Highly classified programs often require personnel with specialized clearances to do the work in special facilities. At the end of production of the aircraft, some OEM personnel can be reassigned to maintenance activity and retain their clearance and expertise on the system, giving CLS an initial advantage. Over the long term, military personnel (who provide most organic squadron-level maintenance) tend to change jobs more frequently than civilian contractors do. Receiving specialized clearances can entail long delays, so maintaining a cleared workforce is more difficult when personnel are frequently reassigned. It may also be difficult to establish separate, highly classified workspaces at government depots that primarily work with unclassified programs.

In some cases, the government has not purchased the technical data, or *data rights*, that provide access to the information needed to perform the sustainment work. This would include, for example, the instructions for making repairs.[3]

GAO reports have commented on this issue. For example, an August 2004 report on PBL observed that "DoD program managers, in contrast, often opt to spend limited acquisition dollars on increased weapon system capability rather than on the rights to technical data," and that this limits their flexibility to compete or develop new sources of repair (GAO, 2004, p. 4). The report included the recommendation that DoD consider requiring program offices to obtain sufficient technical data to allow an alternative source of repair in the future.

A more-recent GAO report noted that the Air Force's C-17, F-22, C-130J, and Airborne Warning and Control System programs had encountered limitations in their sustainment plans because they lacked needed technical data rights (GAO, 2006a). The report reiterated GAO's recommendation that the DoD strengthen its guidance to program managers to address data rights in their acquisition strategies.

Some programs use CLS because of what we term the "acquisition culture" in vogue while the sustainment decision was made. During our interviews, discussants with long careers who had a broad perspective on DoD acquisition and logistics observed that different approaches to

---

[3] Technical data describe or document a product, method, or process acquired by the federal government. In this case, the technical data of interest are repair instructions. The categories of rights to technical data range from unlimited rights to various levels of restrictions, allowing the government to use the data in different ways. Unlimited rights allow the greatest freedom, including the ability to share the data with government organizations or other private vendors. Additional information is available in Nihiser (undated).

weapon system support seemed to come in and out of favor over the last few decades. In the 1980s and earlier, the default source of repair was organic. The exceptions to this rule were commercial items, programs with small fleets, and/or highly classified programs (including programs originally developed for other agencies, including the CIA, such as the U-2). In the 1990s, the Air Force chose CLS for programs that would have had organic support in the past. Stated reasons included

- Political pressure was being applied to move government work to the private sector.
- Acquisition reform, which held that commercial firms offered many benefits unavailable within the government, was a feature of the 1990s. As part of this broader push for change, senior government officials mandated CLS for several programs. Formal policy guidance was to use CLS, with an emphasis on PBL.[4]
- The merger of Systems Command and Logistics Command followed the creation of the program executive office (PEO) structure around 1990. The new PEO structure resulted in a reporting chain from program office to PEO to service acquisition executive (SAF/AQ). This allowed the acquisition community to bypass Air Force Material Command on many decisions or to include it as an afterthought. In addition, the acquisition community was accustomed to dealing with contractors for the development and procurement of weapon systems, and this familiarity continued into the logistics area.

---

[4] Perry, 1994, offered early guidance on acquisition reform. A key element of the reform strategy was greater reliance on commercial products and processes. By 1999, OSD established its Reduction of Total Ownership Cost initiative. A memorandum from the Under Secretary of Defense for Acquisitions and Technology expressed the intention to demonstrate O&S cost savings in pilot programs (Gansler, 1999). The C-17 and F-117 were among the Air Force's pilot programs. Additional policy guidance with an emphasis on privatizing functions not directly related to warfighting followed in DoD, 2001; see especially pp. 53 and 54.

- The end of the Cold War meant decreased resources, and the prevailing belief was that CLS offered more cost-effective logistics support.[5]

Table 5.1 summarizes the reasons interviewees knowledgeable about specific programs gave for using CLS on them. The table indicates only characteristics actually discussed, even if additional characteristics are applicable to the program.

An additional characteristic that may have affected the initial selection of CLS is the origin of the program as an advanced concept technology demonstration (ACTD). Table 5.1 does not include this characteristic because it was not mentioned as a primary reason for using CLS, but it nevertheless probably strongly influenced the decision. The ACTD program began in 1994 with the goal of providing prototype weapon systems to warfighters to use and assess. The emphasis in ACTDs is on rapid demonstration of technology rather than long-term supportability. The Global Hawk and Predator UAVs were developed as ACTDs, and both use CLS.

For Predator, the

> fast pace and relatively short schedule of the ACTD process made it difficult to adequately determine long-term logistics requirements. Similarly, the primary focus of the ACTD was on the demonstration of technology—and the technical performance of the system—and not on how supportable or maintainable the system was. (Thirtle, 1997, p. 45.)

In addition, the government did not own Predator's technical data or maintenance data during the ACTD and did not analyze the type of maintenance that would be optimal for the systems (Thirtle, 1997, pp. 63–64).

Management issues during Global Hawk's ACTD and its transition to a formal development program, such as a lack of involvement

---

[5]   Note that these are the reasons for using CLS that our interviewees advanced. In particular we have no insight into the validity of the third reason listed, although multiple respondents mentioned it.

Table 5.1
Reasons Cited for Use of CLS

Program	Commercial Derivative	Small Fleet	Highly Classified	No Data Rights	Acquisition Culture
B-2		Yes			Yes
C-17	Yes[a]				Yes
DCGS	Yes[b]		Yes	Yes	
E-8	Yes[c]	Yes			
F-117		Yes	Yes	Yes	
F-22				Yes	
KC-10	Yes	Yes			
RC-135		Yes	Yes		
SBIRS	Yes[d]			Yes	
T-1	Yes[e]				
T-6	Yes[f]			Yes	Yes
U-2		Yes	Yes	Yes	

[a] C-17 engines are commercial derivative.

[b] The Distributed Common Ground System has a significant amount of commercial-off-the-shelf (COTS) hardware.

[c] E-8 airframe is commercial derivative.

[d] The Space-Based Infrared System has a significant amount of COTS hardware.

[e] A decision tree analysis showed that CLS would be cheaper for the T-1. The aircraft is a commercial derivative maintained to FAA certification, so its commercial origins probably affected the decision.

[f] The T-6 is a commercial derivative.

of the operational user throughout the ACTD, affected supportability (Drezner and Leonard, 2002, p. 61). During the ACTD, the government did not do significant logistics planning. The investment in such planning was deliberately limited until the system's military utility was assessed. When the program transitioned to production quickly, the lack of planning caused logistics problems (Coale and Guerra, 2006, pp. 8, 9).

In general, ACTD programs do not follow the same acquisition processes as standard procurement efforts. The primary focus of ACTDs is intended to be on demonstrating technology rather than long-term sustainment. Typical sustainment planning was not done for these UAV ACTDs, and considerable system demonstration and procurement took place before a sustainment strategy was developed. During this time, the contractor typically provides the support, and continuing contractor support by default was a natural outgrowth of the demonstration.

In addition to the unique management goals and processes of ACTD programs that led to the supportability challenges in Predator and Global Hawk, the characteristics of UAV ACTDs tend to correlate with three of the reasons shown in Table 5.1. First, ACTDs are supposed to emphasize the integration and assessment of technology rather than its development, so the UAV ACTDs made more use of COTS equipment than aircraft programs that develop new technology would. Second, the UAV ACTDs (particularly Global Hawk) had smaller fleets for many years than are typical of aircraft programs. Third, the government may be less inclined to invest in data rights for a demonstration program with riskier prospects of being mass produced than for a formal development program.

Figure 5.1 illustrates the correlation between fleet size and CLS use for aircraft systems. Further exploration of the outliers provides additional insights. The figure plots budgeted CLS costs per aircraft for FY 2008 against fleet size as measured by primary aircraft authorized. Virtually all aircraft fleets have some CLS funding, even if they are generally regarded as organically supported, if only because of the mandate in AFI 63-111 for CLS support of their training devices. Because most programs use some CLS, it is helpful and more accurate to regard CLS use along a continuum, with the amount of CLS funding per aircraft as an indication of the degree to which the program relies on CLS. Figure 5.1 generally supports the notion that use of CLS is heavier with smaller fleets. The obvious outliers are programs with larger fleets and large amounts of CLS per aircraft; these are labeled in the figure.

The figure generally supports the notion that small fleets are associated with CLS use. Fleets with single-digit sizes—such as those for the C-32, VC-25, C-38, C-40, and E-4 (the five tall bars on the left side of Figure 5.1)—rely heavily on CLS. Conversely, large fleets—such as those for the A-10/OA-10, T-38, KC-135, F-15, and C-130—use little CLS per aircraft. These fleets each total more than 300 aircraft (see the extreme right of Figure 5.1; the CLS funding per aircraft for these is so low that some bars are either barely visible or invisible).

The T-1 and T-6 trainers have fairly large fleets of roughly 150 and 300 aircraft, respectively. Both rely on CLS, but the cost per aircraft is low because these relatively small and simple aircraft are inexpensive to operate and maintain.

Figure 5.1
CLS Use and Fleet Size

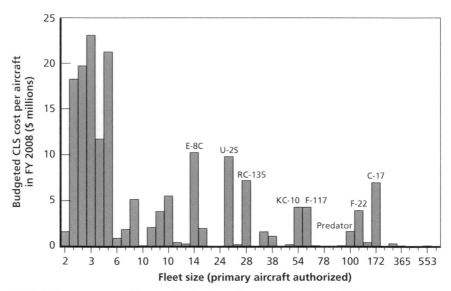

SOURCES: AFI 65-503 Table A7-1 for budgeted CLS costs per aircraft for FY 2008; FY 2008 CLS brochures for the RQ-1, RC-135, and F-117 programs.
RAND MG779-5.1

Figure 5.1 identifies three programs that might be considered out-liers to the relationship just described and another five that clearly are outliers. The E-8 uses a commercial derivative airframe, and its fleet is also fairly small.[6] The U-2 fleet is slightly larger, but the aircraft began as a highly classified program. The RC-135 fleet is slightly larger but still below 30 aircraft and has complex and highly classified equipment.

Five programs are clear outliers to the relationship between fleet size and use of CLS. The KC-10 relies heavily on CLS and has a total inventory of 59 aircraft. It is a commercial derivative that can take advantage of a commercial maintenance infrastructure. The F-117 had a small fleet for a fighter aircraft, plus the additional reasons for use of CLS shown in Table 5.1. The Predator has a fleet size of roughly 100 aircraft and relies heavily on CLS. It began as an ACTD. The most obvious outliers are the C-17 and F-22 programs. The C-17 has some commercial item content, and in both of these programs, issues with data rights and acquisition culture also influenced the decision to use CLS. We will discuss the history of three of the outlying programs, the C-17, F-117, and F-22, more later.

In the 1990s, two programs, the F-117 and the F-22, led the way toward increased use of CLS. The F-117 was originally pursuing an organic support solution at the Sacramento ALC at McClellan AFB. After the BRAC Commission voted in FY 1995 to close McClellan, the Air Force chose CLS for the F-117.

The Sacramento ALC at McClellan AFB would also have been the depot for the F-22. The original sustainment plan for organic support was premised on a fleet of 750 aircraft. When McClellan closed, F-22 sustainment solution was reevaluated. This occurred during the push toward CLS, so that path was chosen for the F-22. A government and weapon system contractor team completed a logistics privatiza-tion study in April 1995.[7] They assumed a fleet size of 442 aircraft and found a slight cost advantage to CLS. The Air Force then chose CLS and used the funds that had been programmed for organic sustain-

---

[6] The E-8 is more commonly known as the Joint Surveillance Target Attack Radar System.

[7] "F/A-22 Logistics Privatization Study," 1995.

ment, including funds to buy data rights, to fix problems in the F-22's development.

The C-17 is the Air Force's largest CLS program and reinforced the trend toward CLS use during this period. Sustainment management and depot repair were planned for San Antonio ALC at Kelly AFB until the BRAC decision in FY 1995 to close the base. With that decision, the Air Force had to rethink its approach. The director of the C-17 System Program Office suggested taking a "flexible sustainment" approach, in which the contractor sustained the C-17 under ICS, and deferring the final choice between organic support and CLS. SAF/AQ and HQ USAF/A4/7 decided to do a cost-benefit analysis of the two alternatives for the C-17. In about 2000, an office was created to collect and analyze data. According to our interviews, the decision was made to pursue partnering to be able to migrate responsibilities back and forth based on the state of the program. The program had started to go in this direction when, in September 2001, an acquisition official in SAF/AQ decided to award Boeing the C-17 CLS as a total system support responsibility contract. The cost comparison was stopped, and the office, and therefore the ability to do this sort of analysis, was closed.[8]

How Are Tasks on CLS Contracts Defined and Funded?

Our insight into CLS contract tasks and funding derives largely from CLS brochures and discussions with program office personnel. The definition of tasks varies among contracts. In some cases, the tasks mirror the elements in the CAIG O&S cost element structure closely. For example, these contracts have tasks with separate line items for aircraft depot maintenance, squadron-level maintenance provided by FSRs, and spare and repair parts. On other contracts, the tasks correspond to parts of the aircraft system, such as logistics support related to the airframe and logistics support related to the engine. Some contracts may use several individual line items for different tasks, and some may list all tasks in one line item to allow maximum flexibility during con-

---

[8] We were unable to find documentation of these events, but more than one of our interviewees told similar stories of the experience.

tract execution. In short, CLS contracts are not structured uniformly with respect to their line items or tasks.

Cost reporting is also not uniform. The level of insight into costs varies according to whether tasks or contracts are fixed price or cost plus and whether the program has negotiated detailed cost reporting. Cost reporting appears, in general, to follow the structure of contract line items, although more detailed cost reporting can be negotiated.

Similarly, the funding structure of CLS contracts also lacks uniformity. We will next consider two aspects of funding structure. One is the extent to which the total amount of funding for a contract is guaranteed for a fiscal year. The other is the freedom to move funds from one task to another within the contract at the discretion of the program manager or contractor.

On some contracts, much or all of the total funding is guaranteed or contractually fixed each year, and the terms of the contract are violated if the contract is not funded to the minimum negotiated amount. On other contracts, little of the anticipated total funding is guaranteed to the contractor each year, and most of the funding is variable at the discretion of the Air Force. On the contracts with more-variable funding, much of the effort is typically tied to aircraft flying hours. So the total funding level of the more-variable contracts is largely at the discretion of the Air Force, and decisions to reduce CLS funding go hand in hand with reduced flying hours.

Our discussions with program office personnel revealed that contractors often offer more-favorable prices in exchange for long-term contracts and/or a guaranteed minimum level of funding. When budgets were stable or increasing, these contracts provided the best value to the Air Force. Now that budgets for maintenance are being squeezed by other priorities, their disadvantages are becoming apparent.

The funding variability written into the contract is important to the corporate Air Force in managing the department's funds for its weapon systems' O&S funds. A program with a large CLS contract with a contractually fixed level of funding may not be able to adapt if the program's own funds are reduced without violating the terms of the contract. Air Force decisionmakers are reluctant to violate a contract's terms and/or renegotiate it while it is being executed because of the potential for disruption. This limits the Air Force's ability to adjust funding among programs, leaving it with inadequate funding for emerging corporate Air Force priorities (readiness of a particular class of aircraft, for example) because it must meet contractual obligations on existing CLS programs.

Table 5.2 illustrates the extent to which CLS contracts include guaranteed or fixed-minimum levels of funding. It shows budgeted amounts for FY 2006 for the ten largest Air Force aircraft CLS programs. The data are based on CLS brochures prepared during FY 2006, so the actual amount of funds obligated in these contracts is different from the estimated amounts shown. Nevertheless, the brochures suggest how the largest CLS contracts are structured, indicating that almost 70 percent of the CLS effort was contractually guaranteed or fixed.

Four of the five largest CLS contracts—the C-17, F-22, U-2, and F-117—included a large amount of fixed or guaranteed funding. Other

Table 5.2
Guaranteed and Variable Funding Amounts in the
Nine Largest CLS Contracts

Contract	Value ($ M)	Guaranteed (percent)	Variable (percent)
C-17	924	69	31
F-22	468	100	0
U-2	325	87	13
KC-10	272	19	81
F-117	244	96	4
RC-135	194	43	57
E-8 JSTARS	153	95	5
C-130J	126	18	82
Predator	105	20	80
JPATS	63	55	45
Total	2,872	69	31

contracts, such as those for the KC-10, C-130J, and Predator, mostly had variable funding, with relatively small amounts guaranteed.

When considering all the elements of weapon system O&S costs, it is important to acknowledge that much of the effort and cost is essentially fixed in a given fiscal year regardless of whether a program uses organic or CLS. In particular, the crew, maintenance, and other personnel and the indirect support associated with the weapon system are largely fixed in the short term. So, for all programs, there is a practical limit to how much funding can be reduced in the short term.

The second sense in which we can consider the structure of CLS contracts is the freedom to move funds among tasks during execution at the discretion of the program manager or contractor. In this sense, CLS contracts again differ across Air Force programs. Some contracts define and fund many line items; for these contracts, changing funding levels among the line items during execution would require a contract action. On the other hand, some program managers deliberately minimize the number of line items to give themselves and their contractors maximum flexibility during execution.

As noted earlier, managers of CLS programs often cited the freedom to move funds among tasks as needed as a benefit of CLS over the restrictions on using funds for organically supported programs.

Why Do CLS Contracts Have So Little Variable Funding?

Our discussions revealed several reasons CLS contracts have so little variable funding. One is that CLS contracts mirror fixed requirements from users. A requirement document might state that a weapon system must be available or mission capable at least 85 percent of the time, for example, and that requirement gets written into the CLS contract. The minimum threshold for availability becomes a minimum threshold of CLS funding to achieve the requirement.

A second reason CLS contracts tend to have so much guaranteed funding is that program managers claim they can obtain more-favorable prices in exchange for offering the contractor more-stable funding. Likewise, CLS providers argue that the assurance of stable funding over several years provides an incentive to invest in such things as scarce technical expertise, data systems, spares pipelines, relationships with vendors, and other resources to provide good logistics support. In times of rising or stable budgets, when funding reductions are not necessary, such arrangements may represent the best value for the Air Force.

A third reason some CLS contracts have little funding variability is that many of the resources required to sustain a weapon system are largely fixed from year to year, especially personnel and facilities. These fixed costs do not vary much with flying hours, and contractors cannot easily make significant changes from year to year. So the invariable nature of the funding reflects the largely fixed-cost structure of the support provided.

A fourth explanation for guaranteed contract funding is that it can protect against funding reductions. A program manager who is asked about how a funding reduction will affect a CLS contract with a large guaranteed funding level can claim that the program is a "must pay" bill and that the terms of the contract will be violated if funding is reduced.[9] While this may be an effective strategy for individual program offices, it reduces the flexibility of the corporate Air Force in meeting budget constraints and imposes a disproportionate share of funding reductions on organically supported programs. It may also squeeze the remaining non-CLS funding on a CLS program, including support provided by the organic Air Force or other contracts.

How Much Insight Does the Government Have Into the Contractor's Costs?

The government's insight into the contractor's costs depends on the level of cost reporting the contract specifies. In general, the Air Force has required little insight into costs on fixed-price CLS tasks or contracts, and more insight on cost-reimbursable efforts.

The two basic types of contracts or tasks are *fixed-price* and *cost reimbursable*. Each type can have several variants. Contracting officials reported that contractors are more likely to agree to provide detailed cost data during the negotiation process on a new contract than for an ongoing contract. For cost-reimbursable contracts, the government can

---

[9] This was confirmed in interviews with program offices.

get monthly reports that break out costs the contractor has incurred and can get detailed information on costs for a component being repaired. The government can get information on how many components were repaired and the cost of each. The Defense Contract Audit Agency can audit these costs. Program offices managing CLS seldom negotiate for similarly detailed costs on fixed-price tasks, according to our interviewees.

An additional factor that affects insight into contractor costs is that some types of contracts are exempt from standard cost-accounting requirements under the Federal Acquisition Regulation (FAR). Contracts or subcontracts awarded on the basis of a firm fixed price or a fixed price with economic price adjustment for commercial items are exempt from cost accounting standards, as are firm-fixed-price contracts or subcontracts awarded on the basis of adequate price competition without submission of cost or pricing data (FAR, 2005, Appendix, Subpart 9903.2). This means that, while the government cannot require certified cost data for such contracts, it can negotiate some level of cost reporting.

During our discussions with Air Force personnel, we heard that contractors often overstate the difficulty of cost reporting and try to negotiate as high a price as possible for such reporting so that the government will be less likely to pay for it. One justification for quoting high prices for cost data is that the requested data are not in the same format the contractor uses, making it more expensive because the contractor would not only have to collect and report the data but also reformat it.[10] Contractors may also be reluctant because they know that reporting costs would also make it possible for the government to calculate how much profit the contractor is making and to use that information in future contract negotiations.

In contrast to its ad hoc collection of CLS costs, the DoD has collected the development and procurement costs of many of its weapon systems in a standardized format for over 40 years in a system called

---

[10] See Lorell, Graser, and Cook, 2005, for a fuller discussion.

Contractor Cost Data Reports.[11] Furthermore, leading companies in industry often require cost data from their suppliers and collect and analyze the data centrally to improve corporate purchasing. For example, Lockheed has created a Strategic Sourcing Solutions group across individual business units (Hannon, 2004), and Honda has a central cost research department that works with the costs of its vendors and shares its expertise throughout the company (Laseter, 1998).

GAO has noted this deficiency in CLS cost data, most pointedly in GAO (2001a), which found, as we did six years later, that "it is impossible to determine whether the cost-effectiveness estimates for proposed CLS approaches are being achieved during implementation because the Air Force does not have the data required to do so" (GAO, 2001a, p. 2).

A related difficulty with cost reporting is that the data are not readily identifiable as depot related. Some of CLS contract costs are funds contractors spend to buy sustainment services at the Air Force depots, which act as subcontractors. Thus, the value of the CLS contracts cannot be used to determine 50-50 calculations without making adjustments for these subcontracts. To comply with the 50-50 law under 10 USC 2466, Air Force personnel try to determine, program by program, which costs are depot related in a process that requires some judgment. GAO has repeatedly criticized the poor quality of cost data for logistics support and the entire DoD process for reporting and forecasting compliance with the 50-50 law.[12]

Official guidance provides for organizations to manage and collect data on logistics support activities. For example, AFI 21-102 provides guidance on depot maintenance management policy and identifies AFMC as responsible for management and execution of the Air Force depot maintenance program. AFMC determines core maintenance capabilities and compliance with the 50-50 law and, with the program offices, conducts source-of-repair analyses. AFI 21-133(I) is

---

[11] This is not true for programs conducted under "price-based acquisition," one category of acquisition reform. Lorell et al. (2005) describe the issues with PBA.

[12] The most recent GAO report we found on the subject was dated November 2006. Appendix B summarizes other, earlier reports on cost data and 50-50 reporting

the Air Force instruction on the DoD-wide Joint Depot Maintenance Activities Group. The group is charged with developing and maintaining Depot Maintenance Operations Indicators and the Depot Maintenance Cost Comparability Handbook. Together with service representatives, the group is instructed to ensure that cost data provided by public and/or private activities submitting proposals to perform depot maintenance during competitions are analyzed in a consistent manner.

Although the guidance for the capability to gather and analyze cost and performance data on logistics support exists, data on contractor logistics organizations are not being systematically collected in the Air Force and made available to those who manage CLS contracts. Similarly, few people in the Air Force have access to information on public depots.

How Are CLS Contract Prices Determined?

CLS contracts and tasks for major weapon systems are generally not awarded under competition. When competition is available, it is most often for commercially derived systems, such as trainers and operational support aircraft, when private organizations already provide support for the commercial item. These programs represent a small fraction of Air Force O&S costs. Competition is usually difficult or impossible on most programs, because the OEM has such a significant advantage in providing logistics support for a system that it already produces and probably already supports (through ICS) or because the government does not own data rights that would allow it or other organizations to support the system.

The initial price on noncompetitive fixed-price tasks is determined through a process of negotiation between the government and the contractor. When the program office gets a proposal from the contractor, government engineers review the proposed labor hours per task. They use analogies, their knowledge of the work to be done, and other tools to reach agreement with the contractors. The engineers may know the actual hours incurred on previous, similar work. Then, the Defense Contract Audit Agency and the Defense Contract Management Agency audit the proposed cost per labor hour, looking at the price ranges and structures. These agencies do not look at hours but at requirements and certify that the rates are reasonable. In this way, the government increases its confidence that it is getting fair prices.

For cost-reimbursable tasks or contracts, the contractor is reimbursed for its costs plus a fee or profit. The government audits the contractor's costs to ensure that they are charging for only the actual costs incurred to perform the specified work.

Observations on CLS Management

Interviews suggest that several weapon system program characteristics are associated with CLS use, including whether the program uses commercial derivative or COTS equipment, the fleet is small, the programs or technology is highly classified, and whether the government lacks the technical data or data rights. In the few analyses supporting source-of-repair decisions that we have seen, most of these characteristics have been decisive. Another CLS association is whether the program originated as an ACTD. The analyses required in the Air Force's source-of-repair analysis process should determine whether these and other characteristics influence the decision to use CLS.

Most of the largest CLS contracts are structured to provide high levels of guaranteed funding each year. The Air Force now finds itself with increasing pressures on O&S budgets and is interested in moving away from guaranteed funding levels for large CLS contracts.

The Air Force does not collect detailed and uniformly formatted costs for CLS as it does for organic O&S costs and for large development and procurement contracts. The lack of data makes it difficult or impossible to determine whether prices are reasonable and to estimate future costs for budgeting, contract negotiating, and other purposes. The lack of uniform standards for CLS cost reporting is in marked contrast to the requirements to report DoD development and procurement contract costs in a prescribed cost-element structure.

Few CLS contracts are awarded through competition, which means that contractor selection and price determination require analysis and negotiation. To ensure that the Air Force is selecting the best source of repair and is obtaining reasonable prices, such analysis requires good cost and performance information, which the Air Force does not systematically collect, as well as program office personnel experienced in CLS contracting, cost and price analysis, and other business skills.

CHAPTER SIX

Implications for Cost Analysts

As we stated at the outset, this monograph has two purposes: assessing CLS use in the Air Force and providing insights for cost analysts addressing Air Force O&S costs. This chapter describes the six main implications of CLS for cost analysts that emerged from our research:

1. Funding sources may shift at different stages of the weapon system's support, and the visibility of the funding may change accordingly.
2. CLS affects the amounts and proportions of costs reported in non-CLS O&S elements in AFTOC.
3. The nature and scope of CLS tasks differ among programs.
4. We found no clear evidence of differences in cost or rates of cost growth between organic support and CLS.
5. Some CLS costs are accounted for differently than are the corresponding organic costs.
6. It is difficult to generate cost-estimation relationships for total system O&S costs because much of the total cost is affected by funding constraints. The implications are addressed in more detail later.

Funding Sources May Shift at Different Stages of Support

Weapon systems typically are supported by the prime contractor during production under ICS and then change to permanent contractor or organic support. The support and funding provided under ICS may not be reported in AFTOC, which may therefore give an incomplete picture of these O&S costs. Analysts can identify ICS funding in the Aircraft Procurement, Air Force budget exhibit P-5. Logistics support funded by procurement appropriations may have different names for different programs. For example, for the F-22 program, it appears in the P-5 exhibit as Performance-Based Agile Logistics Support. For the C-17 and C-130J programs, it appears as ICS. Unfortunately, the budget exhibits provide no further details about the nature of the tasks, but cost analysts should at least be aware of the amount of ICS funding provided with procurement funds each year and that the O&S costs reported in AFTOC for ICS-supported programs do not represent all the logistics support.

CLS Affects the Amounts and Proportions of Costs Reported in Non-CLS O&S Elements

As mentioned earlier, CLS costs are reported as a lump sum in AFTOC under the CAIG O&S cost element for CLS because CLS costs are not collected in a uniform format or reported in detail to AFTOC. CLS tasks can span all the CAIG elements that organic support also performs, even though AFTOC reports the costs under the CLS element. This inability to report CLS costs in the appropriate element means that cost analysts do not know the true costs of each element, such as consumables, DLRs, and aircraft overhaul, because only the organic costs are reported in the correct element. So, CLS can distort the explicitly reported cost of O&S elements and their proportions of the total. This potential distortion is particularly important when comparing the costs or the proportions of the costs of individual elements across programs.

The Nature and Scope of CLS Tasks Differ Among Programs

The CLS brochures that weapon system program offices prepare describe each CLS task and its budgeted cost. The brochures provide estimates for the year in which the brochures are submitted and for the next several years. Thus, the brochures do not reflect actual obligations for the current year and can deviate from actual obligations, depending on what happens after the brochure is submitted. Although the estimates do not represent actual costs, the brochures provide good insight into the nature of the CLS tasks and a generally good estimate of their cost for the current year. The CLS brochures we reviewed showed that the nature and scope of CLS support varied widely, depending on the type of weapon systems and on the aircraft system. We found that space systems use proportionally more CLS than do aircraft systems and that most space system support is sustaining engineering.

In contrast, most CLS for aircraft systems consists of depot maintenance and the repair and replenishment of parts. But the scope of the CLS varies a great deal among different aircraft systems. For some aircraft programs, most of the CLS may be for a single subsystem, such as a radar, or for a single element of support, such as aircraft overhaul. For other aircraft programs, CLS can provide the majority of the logistics support and span every element of support. Again, the nature or scope of the support is not evident from AFTOC but must be identified from another source, such as the CLS brochure.

Cost and Cost Growth

Because we lacked detailed data, we found no clear evidence of cost or cost growth differences between organic support and CLS over time. This finding may be useful to cost estimators who are using costs of an analogous weapon system to estimate the cost of a new system using a different support concept. The same work is rarely done on the same program or piece of equipment by both the government and a contractor, something that would offer the best evidence for cost comparisons. However, we found data on only one such case. For comparing cost growth, we tried to compare organic and CLS costs for similar types of aircraft programs that had fairly stable inventories and usage rates over the last several years. Again, it was difficult to find comparable cases that met these conditions, and the evidence on cost growth in these few cases was mixed.

Based on the lack of clear evidence either way, cost analysts who are estimating O&S costs and cost growth should treat organically supported programs and CLS programs the same.

Some CLS Costs Are Accounted for Differently Than Are the Corresponding Organic Costs

A major difference that we have seen in the accounting of O&S costs is accounting for the personnel who manage the supply chain, for example, item managers, logisticians, and sustainment engineers. For organic programs, the cost of these personnel is part of the surcharge added to costs of consumables and DLRs. What appears in AFTOC as the cost of consumable and reparable parts actually includes the overhead cost of the personnel who manage the supply chain of the parts and repairs. For CLS programs, contract line items tend to report material and personnel costs separately. Because of this accounting difference, the cost of sustaining engineering or similar personnel-intensive tasks might appear disproportionately high for CLS programs relative to organically supported programs, and the cost of consumable and reparable parts may appear lower than if overhead costs were included, as they are in the accounting of organic consumable and reparable costs.

It Is Difficult to Generate Cost-Estimating Relationships for Total System O&S Costs Because Funding Constraints Affect Much of the Total Cost

Estimators working with O&S costs sometimes estimate costs closely associated with flying hours, such as unit-level consumption and depot maintenance costs, and sometimes estimate total O&S costs, which include costs more loosely associated with flying hours. A common way to handle costs that vary closely with flying hours is to estimate them on a cost per flying hour basis. Total O&S costs are also often estimated on a cost per flying hour basis in recognition of the important role that flying hours have on total O&S costs.

However, estimators should understand that O&S costs, particularly those less closely related to flying hours, are affected by the amount of funding available in a given year. Maintenance manpower costs, for example, may be constrained by the availability of qualified maintenance personnel or by funding. Similarly, sustainment support costs, such as for modifications and sustaining engineering, may be deferred or reduced because of funding constraints. These elements of O&S that contribute to total O&S cost are therefore likely to vary with available funding and not with flying hours. This variation in total O&S cost makes it difficult to develop cost-estimating relationships that accurately express O&S costs as a function of known variables, such as aircraft size, mission, or usage rates, because of the unknown variation in funding constraints.

CHAPTER SEVEN

Summary and Recommendations

This chapter summarizes some of the main points from Chapters Three, Four, and Five to set the stage for discussion of five recommendations about CLS use. The five are linked with problems we identified in the course of the study through interviews with people knowledgeable about CLS in the Air Force, most often as part of the Air Force's own CLS IPT, and in other government organizations, as well as contractors themselves. The Air Force is already taking steps to implement changes that address at least three of the recommendations.

Summary of Findings

Chapter Three provided an overview of CLS funding in the Air Force. The growth in CLS costs in the Air Force has been driven largely by growth in the inventory of aircraft supported by CLS. Legacy aircraft tend to be organically supported, and newer aircraft tend to be CLS. The two largest CLS contracts in FY 2006 illustrate this trend: The C-17 replaced the organically supported C-141, and the ongoing F-22 replaces legacy fighters that are organically supported.

Chapter Four addressed CLS performance. The anecdotal evidence on CLS performance among the Air Force and other knowledgeable people who had experience with CLS varied widely. We are concerned about the lack of solid cost or performance data to support conclusions about CLS performance, especially with regard to depot performance. Although we were able to compare supply-system performance as measured by total not mission capable–supply rates, this analysis was not conclusive because of uncertainties about levels of funding and other factors affecting supply-system performance for organic and CLS programs.

Chapter Five addressed the CLS management process. When new programs begin the source-of-repair process, the program offices, with the assistance of AFMC/A4, make decisions on a case-by-case basis. The source of repair is seldom determined by competition but usually by analysis. Similarly, contract prices are usually determined by analysis and negotiation rather than competition. Few programs are able to use competition, sometimes because the government has decided not to buy technical data rights that are necessary to maintain complex weapon systems. Using analysis rather than competition to make decisions on sources of repair and to determine contract prices underscores the importance of the quality of the data used in the analysis.

CLS costs are not reported to program offices in detail or uniformly. The level of detail and its structure are left to the discretion of the program manager. Furthermore, the cost data, other than the total contract value, are not normalized or reported beyond the program office. The lack of data made it impossible to assess CLS cost performance confidently, including whether initial costs are reasonable or whether ongoing costs have been increasing faster than organic costs.

CLS contracts often include high annual levels of contractually guaranteed, or fixed, funding. Program office personnel reported that they were able to negotiate better prices in exchange for guaranteed funding levels. While such arrangements made sense when budgets were stable or increasing, they have reduced the flexibility of the corporate Air Force to apportion funding reductions, leaving organically supported programs to bear a disproportionate burden of budget cuts. From the program manager's perspective, guaranteed minimum funding levels offer some protection against funding cuts because the program manager can claim that CLS is a "must pay" bill.

Recommendations

Several of the problems we have discussed involve decentralized decisionmaking by program offices. Program offices sometimes make decisions that are in the best interest of the program but not in the best interest of the Air Force as a whole. Thus, several of the issues involve either increasing centralization of decisionmaking or placing stronger constraints or requirements on program offices. The Air Force's CLS IPT, in which we participated, was formed to address the problems regarding cost data and funding flexibility discussed here. The Air Force is in the process of implementing changes to address these problems. In addition, the Air Force has drafted guidance to address the issue of buying technical data.

Require Centralized Decisions on Buying Design and Technical Data or Use Rights to Data

An organization that maintains a weapon system needs the technical data from the original manufacturer to maintain the system. We heard of numerous instances in which the Air Force failed to buy technical data, or even the usage rights to the data, early in the program and was subsequently unable to obtain the data at an affordable price. In some cases, the original manufacturer refused to sell the data; in other cases, the price was so high that the program manager believed it was unaffordable within the constraints of the program. In either case, the lack of technical data means that the Air Force will be unable to maintain the weapon system itself or to hold an effective competition for maintenance, and the original manufacturer becomes the de facto sole source of maintenance. In these situations, the Air Force has little recourse when dissatisfied with the cost or quality of the contractor's work.

In recent years, the Air Force and Congress have addressed this issue for future sustainment programs. The Deputy Assistant Secretary of the Air Force, Contracting, addressed the issue with advisory memoranda dated March 27, 2001, and February 11, 2002. The first memo clarified the position of the FAR on the purchase of technical data. The second memo urged contracting officials in the field to consider data rights early in the acquisition process and offered resources to assist them in doing so.

More recently, the FY 2007 National Defense Authorization Act revised 10 USC 2320 to require DoD program managers of major weapon systems to "assess the long-term technical data needs of such systems and subsystems and establish corresponding acquisition strate-gies. . . . ." The revised law falls short of requiring program managers to buy technical data.

Program managers typically give up data rights in favor of fund-ing a near-term developmental need, but they will still face these issues. Buying the technical data or data rights early in the program, at the competitive stage, would discourage competing contractors from charging exorbitant prices for the data, because it could reduce their chances of winning the initial contract for the program.

Because program managers do not necessarily have the same incentives as the corporate Air Force, a centralized Air Force office should monitor compliance with this law and make the final deter-mination on buying technical data. (SAF/AQ and HQ USAF/A4/7 should both have input in this decision.)

Guidance in proposed revision of AFI 63-101 (draft, 2008) imple-ments the provisions of the revised 10 USC 2320 about data rights and addresses this issue. The draft AFI directs the program manager to assess requirements for data rights over the life of the weapon system, and address the subject at ASPs and reviews.

In addition to monitoring and making the final decisions about buying technical data, the corporate Air Force could

- direct managers of programs that do use CLS but do not have the technical data needed for another source of repair to determine the cost of obtaining such data
- direct program managers to work in conjunction with AFMC to determine when competition for logistics services may be beneficial
- consider the above results in making case-by-case determinations about purchasing technical data.

We expect this draft guidance, if implemented, to position the Air Force to be able to retain more choices for the logistics support for its weapon systems and motivate better performance.

One challenge in implementing this guidance is that, even under existing guidance, the Air Force has convened ASPs and similar corporate-level reviews of sustainment plans, including plans regarding technical data, and has failed in many cases to purchase the technical data or data rights. Perhaps the congressional attention shown in the 2007 revision to 10 USC 2320 will focus the attention of the corporate Air Force on the subject.

Require a Uniform Format for Cost Data

Cost data on CLS contracts are not currently reported in a uniform format. The structure of the cost data and the level of detail reported vary from program to program. Some programs supply data identifying the major types of O&S costs, such as DLRs, aircraft overhaul, engine overhaul, or sustaining engineering, but the data others supply do not allow identification of costs at this level of detail. Regardless of the level of detail the program office collects, only the lump-sum value of the CLS contract makes it into AFTOC, thus giving analysts using the Air Force's official O&S cost database little insight. Analysts working with AFTOC data are unable to determine which CLS costs vary with usage rates, such as DLRs, and which are relatively insensitive to usage rates, such as sustaining engineering.

We recommend that SAF/AQ and/or SAF/FM require program managers to collect and report CLS and ICS costs in the standard CAIG O&S cost-element structure. The data need not be certified costs as defined in the FAR. The costs should be reported at least to the second level of the work breakdown structure, for example, differentiating between consumable and reparable parts and between aircraft and engine depot maintenance. Furthermore, the Air Force should consider clearly identifying costs incurred at Air Force and contractor depots as such. Alternatively, AFMC data systems could be improved to record the organic depot expenditures that are associated with subcontracted workloads from CLS or other contractor-organic partnerships. Finally, the costs should be reported in an electronic format that is compatible with Air Force cost databases so that the detailed CLS costs can be easily incorporated into AFTOC and other Air Force databases, and the Air Force should ensure that the data are retained corporately and are accessible by other CLS program managers.

Some program office personnel who have been asked to report current CLS costs in the standard O&S format have argued that the categories do not always match their CLS tasks. This objection seems to stem from a lack of familiarity with the cost format, however. The cost structure adequately captures organic O&S tasks and costs. Because CLS is defined as the provision of the same O&S elements as organic support, the same reporting structure should, by definition, adequately capture the tasks and costs provided by either source.

Uniform CLS and ICS cost reporting should have a few significant advantages. First, using the standard O&S cost elements would allow comparison of costs across weapon systems. Second, the uniformity would help identify which costs vary with usage rates and which are relatively insensitive to them. These comparisons and insights would be useful for assessing the effects of changes in aircraft flying hours on CLS funding. Third, the recommendation would result in a system of cost reporting that would enable better compliance with the 50-50 law. Fourth, standardized, reasonably detailed, and easily transferable CLS and ICS costs could contribute to a repository of logistics support providers' costs that could inform Air Force contract negotiations and improve its estimates of O&S costs for future systems. The fourth advantage bears some amplification.

As discussed earlier, competition is seldom used to determine the repair provider or to set prices for logistics services. By necessity, then, prices are usually determined by analysis and negotiation. The quality of the analysis depends on the quality of the cost data. CLS and ICS cost data lack detail and a uniform structure, and the Air Force does not collect them centrally. As a result, the Air Force lacks good data on which to base its selection of repair providers, its pricing analysis, and its overall approach to logistics support. Mandatory reporting and collection of well-defined ICS and CLS costs across the Air Force enterprise would begin to allow the Air Force to make more informed logistics decisions.

The Air Force has drafted guidance (AFMC, 2008) requiring that requirements for CLS funding be submitted in the standard CAIG cost-element structure as part of the Centralized Asset Management process. This is a step in the right direction that should be taken further.

Provide Centralized Guidance to Achieve Flexibility

As we have discussed, most large CLS contracts are written so that most if not all the funding each year is guaranteed to the contractor or is contractually fixed. The terms of the contract are violated if the Air Force does not provide the guaranteed level of funding. At the other extreme, a few large CLS contracts are written so that most of the effort is contractually variable or is determined by the government, with relatively little of it being guaranteed each year.

From the program manager's perspective, financially inflexible contracts make sense for several reasons. First, contracts that specify a certain amount of funding to achieve a given level of operational availability tend to reflect the inflexible operational requirements given to program managers. Second, program managers can offer a guaranteed level of funding in exchange for more-favorable pricing. In times of rising or stable budgets, this arrangement works well for the Air Force. Third, many of the elements of sustaining a weapon system, especially personnel and facilities, are difficult to change significantly from year to year. And fourth, when facing budget cuts, program managers with large amounts of contractually guaranteed funding can plead that they have bills they must pay and therefore cannot accept any funding reductions. When budgets are not rising or stable but are declining, this arrangement works poorly for the Air Force as a whole because disproportionate reductions must be made in organically supported programs and the more-flexible CLS programs to avoid breaching CLS contracts that have guaranteed funding levels. Disproportionate budget cuts may occur even when they run counter to the corporate interests of the Air Force or the goal of meeting the 50-50 law.

Program managers might argue that they are simply writing into the terms of the contract the requirements of HQ USAF or the operating commands. Requirements are usually inflexible and demand a certain minimum level of operational availability or flying hours. A reasonable counterargument is that organically supported programs do not enjoy the same kind of funding guarantee to achieve requirements and that Air Force leadership must retain the flexibility to manage funding and readiness among all its programs. While addressing the problem should optimally involve those who set the requirement, the sustainment community should be able to independently structure more-flexible contracts using currently available contracting structures.

The Air Force now finds itself with intense demands on its O&S budgets and is interested in moving away from guaranteed funding levels for large CLS contracts. The Air Force has issued a memorandum advising program managers to create more-flexible CLS contracts (Assistant Secretary of the Air Force, Acquisition, 2008) . In addition, the draft revision of AFI 63-101 addresses flexibility and directs program managers to work with contractors, users, and other agencies to establish flexible performance and funding ranges on CLS contracts.

Flexible CLS contracts, in conjunction with cost reporting that allows decisionmakers to see what is bought through CLS, should allow the corporate Air Force to allocate O&S funding more intelligently among programs during the budget and execution process.

Achieving greater flexibility in the funding of CLS contracts may also have a couple of drawbacks, however. First, it may increase prices for CLS contracts at a given level of performance because contractors are being asked to assume more funding risks, although the Air Force may judge the increased flexibility is worthwhile anyway. Second, flexible funding could disrupt incentives on some performance-based contracts that are structured to motivate contractor performance at a set funding level, although presumably the contracts could be rewritten to allow incentives at various funding levels.

Strengthen Centralized Expertise to Optimize CLS Use

Among our more interesting findings is the wide divergence of opinion on the cost and effectiveness of CLS. Most but not all people in program offices that used CLS thought it was better than organic support. Others believed differently.

Most strikingly, few people could offer evidence to support their positions. Because there is so little competition for logistics support, there are few instances of comparable work being performed that would provide compelling proof. Of equal concern, there appears to be no systematic data collection on CLS cost or performance that would allow the Air Force to begin to address the question. Furthermore, we were not given access to cost or performance metrics on Air Force depots, although we were assured that such data are collected. It appears that few people in the Air Force have access to such data, and only a handful of people in individual program offices have access to performance data on the CLS contracts for their programs. Given the difficulty of making accurate comparisons and the sensitivity of the results for both contractors and organic depots, the lack of open access to such data is unsurprising.

Similar to the lack of relevant data to support decisions on CLS, the Air Force suffers from a lack of human capital. A program office holding a source selection for a CLS contractor or writing a new CLS or PBL contract or modifying an existing one needs expertise in these areas in addition to data. The Air Force's expertise in these areas is scattered in various places, and one complaint we heard from some program office personnel was the difficulty of getting personnel with the right expertise to help develop performance-based metrics and to help manage other aspects of CLS.

Decisions on sources of repair that affect compliance with the 50-50 and core requirement laws also appear to be decentralized. AFMC/A4B makes the initial determinations about source-of-repair decisions that affect compliance with both laws, and the Acquisition Strategy Panel considers and approves the overall logistics support plan early in a program's life cycle. However, because so little is typically known early in a program, the nature and amount of work may change over time. The program office makes the work allocations that affect 50-50 and core requirements, theoretically in conjunction with AFMC, although the degree of collaboration varies in practice. In addition, workloads may shift over time from the original plans. Recent and significant examples include the transition from contractor to organic performance for F100, F117, and F119 engine depot maintenance and for KC-135 PDM and C-17 airframe and nonairframe depot maintenance. These workloads were brought to organic sources to allow the Air Force to comply with 50-50 requirements.

The Air Force has expertise in logistics support, including CLS, in several headquarters and field-level organizations. Headquarters organizations include the Centralized Asset Management office in AFMC/A4, which oversees the requirements for CLS and organic depot maintenance; USAF A4/7; SAF/AQ; and the Acquisition Center of Excellence offices. The expertise found in these is marshaled as needed, such as when an acquisition strategy panel is convened to approve the LCMP for an individual program. However, this expertise is not sustained in a coordinated, ongoing effort that examines the enterprisewide effects of sustainment decisions on individual programs. Further, the data needed to consider and optimize sustainment decisions are not collected across the Air Force enterprise. For example, we have been unable to identify an office in the Air Force that tracks the costs of standing up a depot repair capability, often a major factor in decisions regarding where such work should be performed.

We recommend that the Air Force synthesize and increase the expertise needed to manage logistics support. This might be done by establishing or strengthening an existing centralized office and/or by enhancing a career field, such as acquisition logistics, and ensuring that personnel in the field are highly trained and prominent throughout the acquisition chain of command. Success in this area requires the Air Force to

- collect and analyze cost and performance information about both organic and CLS providers
- develop and offer expertise in CLS including cost, logistics management, and PBL to program offices
- monitor and participate in ongoing decisions that affect compliance with core and 50-50 laws
- assess the effects of decisions on the sustainment of individual programs on the sustainment capabilities of the Air Force enterprise as a whole.

Successful implementation of this recommendation would require writing CLS contracts that require standardized cost data and requiring program offices to collect and report relevant performance data.

We recognize that there may be political issues with making data available that might make organic depots look worse than contractors or vice versa. However, Congress has made it clear that the 50-50 regulation was written in part to maintain an industrial capability in case of war. This reasoning would still hold no matter how well or how poorly the depots performed. More widely available data on cost and performance would also presumably spur both contractors and organic depots toward improved efficiency.

Retain Choices for Logistics Services
Few CLS contracts are competed. Competition is often impossible because the government does not own the data rights for systems to be maintained by organic depots or third-party providers. In other cases, the incumbent may have an advantage over potential entrants. Program office personnel told us that contractors are often unwilling to bid on CLS contracts unless the contract terms are long enough to make it worthwhile for them to purchase the required spares and the initial nonrecurring tooling required to do the work. The absence of competition for CLS means that initial and ongoing prices on CLS contracts are seldom affected by competition. Program managers often try to motivate better performance through the use of award fees, incentive fees, or extended contract terms. These measures can be effective in motivating performance, but we also heard program office personnel complain of an unresponsive contractor on a large CLS contract despite such incentives. The program office lacked data rights on this program, and the contractor enjoyed a monopoly.

Such measures as award or incentive fees are typically used to motivate performance but not to motivate price reduction. Cost-reimbursable tasks may offer little incentive for contractors to reduce costs, except as a means to help them win future contracts if a viable competitor exists. On fixed-price tasks, program offices generally rely on government personnel to assess whether proposed labor hours are reasonable. But without information from other organizations on comparable tasks, it is difficult to assess the reasonableness of the offer.

We recommend that the Air Force enhance its ability to choose among viable options when looking for providers for logistics services.

This recommendation must be enabled by three of the preceding recommendations. The ability to choose among providers requires that the government have

- data rights, so that organizations other than the OEM have access to the technical data needed to maintain and repair the equipment
- a history of cost and performance data for comparable work, including work on other programs, so that it can assess the prior performance of support providers and the reasonableness of proposed prices
- a cadre of experienced contracts, business, and logistics personnel to negotiate and manage contracts.

Retaining choices among logistics providers may require assessing logistics support at the commodity or subsystem, rather than the platform, level. A contractor may have an advantage in providing logistics support for an engine, landing gear, or certain electronics equipment but not for an entire platform. The Air Force may need to assess and make source-of-repair decisions below the platform level to determine the benefit of outsourcing everything to a single CLS contractor rather than outsourcing at the subsystem level to multiple sources of repair.[1]

The primary benefits of retaining choices for logistics support would be better prices and performance. While we found no evidence that CLS costs are growing more rapidly than organic costs for comparable systems or that CLS performance lags that of organic logistics support, public-sector supply system and depot costs have grown far faster than the general rate of inflation. Perhaps the appropriate metric for assessing CLS performance should not be "as good as organic support" but rather "as good as a more-open market will allow."

---

[1] GAO (2004) supports this recommendation, suggesting that PBL contracts on a platform or system level are rarely if ever used in the commercial world. However, there are trade-offs to consider, since this approach could potentially reduce contractor accountability for system-level cost and performance and would require the Air Force to integrate and manage a larger number of individual contracts.

Comparison of Supply-System Performance on CLS and Organic Programs

All the largest aircraft CLS programs provide supply support. The terminology varies somewhat, with the most common descriptions being contractor operated and maintained base supply, supply support, inventory control point, and supply chain management. We made two kinds of comparisons of supply-system performance. One comparison measures the Air Force standard for TNMCS for each aircraft against the achieved rate over the three-year period from the second quarter of FY 2003 through the second quarter of FY 2006.[1] The achieved rate represents the average of the quarterly rates over the three-year period. The second comparison is of the achieved TNMCS rate between CLS and organic aircraft with the same mission. The comparisons are shown in Figures A.1–A.4. The solid bar on the left of each pair is the achieved TNMCS rate, and the hatched bar on the right of each pair is the TNMCS rate standard for the aircraft indicated. The lower the TNMCS rate, the better. While most aircraft, whether contractor or organically supported, meet their standards, CLS aircraft are held to tighter standards than their organic counterparts. This could be because the CLS programs are funded more generously to achieve a more-demanding standard, rather than because of the superiority of CLS under equal conditions.

---

[1] The Air Force sets a standard or goal for TNMCS for each aircraft mission, design, or series combination that is a function of the aircraft availability target or CLS goal for that aircraft.

Figure A.1 shows TNMCS rates relative to Air Force standards for selected trainer aircraft. The T-1 and T-6 are CLS aircraft and the T-37 and T-38 (the rates for the T-38s at Air Education and Training Command and Air Combat Command are reported separately in MERLIN) are organically supported, except that the T-38Cs have CLS for equipment that is unique to the C variant. All the trainer aircraft met the relevant Air Force standards, although the standards for the CLS programs are much higher.

Figure A.1
TNMCS Rates and Standards, Selected Trainer Aircraft

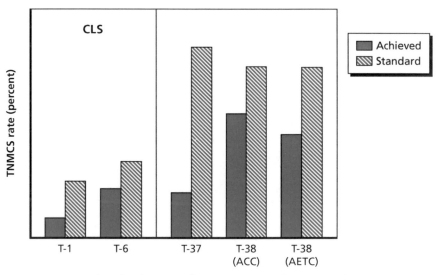

SOURCE: MERLIN data for the active fleet, average rate from second quarter FY 2003 through second quarter FY 2006, October 2006.
RAND MG779-A.1

Figure A.2 shows achieved TNMCS and the standard for active-duty cargo aircraft, including the C-17 ICS/CLS aircraft and organically supported cargo aircraft. The C-17 has a more-demanding standard, and its achieved performance was better, both absolutely and relatively, than that of the organically supported aircraft. The C-130E and C-130H aircraft met their Air Force standards. The C-5 does not but has always been notorious for reliability problems.

Figure A.2
TNMCS Rates and Standards, Selected Cargo Aircraft

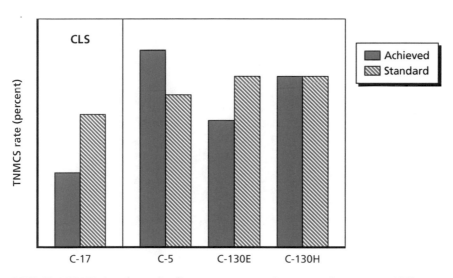

SOURCE: MERLIN data for active fleet, average rate from second quarter FY 2003 through second quarter FY 2006, October 2006.
RAND *MG779-A.2*

Figure A.3 shows achieved and standard TNMCS rates for active-duty fighter aircraft, including the F-117 CLS aircraft and selected organically supported fighters. The F-117 program had two unusually bad quarters during this period, which pushed its average achieved rate above its more-demanding standard. It still performed better than the organic fighters, except that its TNMCS rate was nearly equal to that of the F-16A/B. All the organically supported aircraft, except the F-15C/D, met their standards, which are looser than those for the F-117.

Figure A.3
TNMCS Rates and Standards, Selected Fighter Aircraft

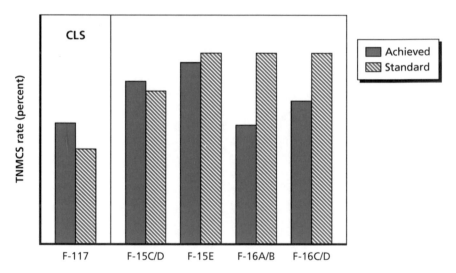

SOURCE: MERLIN data for active fleet, average rate from second quarter FY 2003 through second quarter FY 2006, dated October 2006.
RAND MG779-A.3

Figure A.4 shows achieved and standard TNMCS rates for active-duty tanker aircraft, including the KC-10 CLS tanker aircraft and the organically supported KC-135R/T variants. Both aircraft met their standards, although, once again, the CLS aircraft has higher standards than does the organic aircraft. Note that the KC-10 has an average age of roughly 22 years; and the KC-135 is twice as old, and the KC-10 has significantly lower break rates.

Figure A.4
TNMCS Rates and Standards, Tanker Aircraft

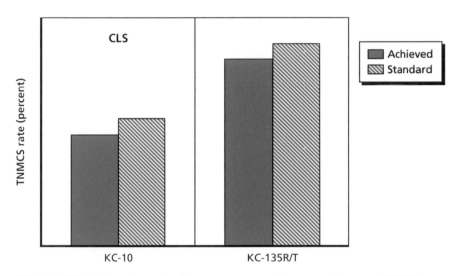

SOURCE: MERLIN data for active fleet, average rate from second quarter FY 2003 through second quarter FY 2006, dated October 2006.
RAND MG779-A.4

In these comparisons, the standards for CLS programs are always more challenging. The cargo and tanker programs supported by CLS met their standards more successfully than did their organic counterparts, while the supply-system performance of fighters and trainers was more ambiguous. Because the TNMCS metric depends so much on funding for spare parts, as well as other factors, we cannot say that this measure of performance indicates better supply-chain management—it could be simply that programs with better TNMCS have received better funding.

Laws, Directives, Regulations, Instructions, and Reports That Affect CLS Use

This appendix is divided into five parts: a brief summary of the congressional legislation related to depot maintenance from the last 50 years or so; a brief history of the reports, testimony, and findings of GAO (the investigative arm of Congress); the governing OSD regulations; relevant DoD IG reports; and finally, the Air Force directives related to depot maintenance activities.

Legislation in Title 10 of the U.S. Code

This section briefly summarizes the sections of Title 10 (listed in numerical order) related to depot-level maintenance and repair. The major themes of these laws are defining what depot maintenance activities are; ensuring that a wartime depot maintenance capability under the control of DoD will be available; maintaining a robust organic capability (called a "core logistics capability") that could expand to meet wartime requirements; and providing depot maintenance services efficiently to military customers through the use of competition, when appropriate. These sections of U.S. Code provide complete details.

10 USC 2208(j), Working Capital Funds

This section permits DoD industrial facilities funded by a working capital fund to manufacture or remanufacture articles, as well as to provide manufacturing and engineering services and sell them to customers outside DoD.

10 USC 2320, Rights in Technical Data (as amended by the National Defense Authorization Act for Fiscal Year 2007)

This section addresses the government's rights to technical data for items and processes. The 2007 amendment requires

> program managers for major weapon systems and subsystems of major weapon systems to assess the long-term technical data needs of such systems and subsystems and establish corresponding acquisition strategies that provide for technical data rights needed to sustain such systems and subsystems over their life cycle.

The assessment is to be done before contract award and is to consider priced contract options for the future delivery of technical data.

10 USC 2460, Definition of Depot-Level Maintenance and Repair

This section defines depot-level maintenance and repair as activities requiring the overhaul, upgrading, or rebuilding of parts, assemblies, or subassemblies, and the testing and reclamation of equipment as necessary, regardless of the source of funds for the maintenance or repair or the location at which the maintenance or repair is performed. The term includes (1) all aspects of software maintenance classified by DoD as of July 1, 1995, as depot-level maintenance and repair, and (2) ICS or CLS (or any similar contractor support), to the extent that such support is for the performance of services described in the preceding sentence.

Depot-level maintenance and repair does not include major modifications or upgrades of weapon systems that improve program performance or the nuclear refueling of an aircraft carrier. Private or public-sector activities would continue to perform major upgrade programs covered by this exception. The term also excludes the procurement of parts for safety modifications but does include their installation.

10 USC 2462, Contracting for Certain Supplies and Services Required When Cost Is Lower

This section directs the Secretary of Defense to procure each supply or service necessary to accomplish the authorized functions from a source in the private sector if it can provide the supply or service at a lower cost than DoD can provide it, unless the Secretary of Defense determines the function must be performed by military or government personnel.

10 USC 2464, Core Logistics Capabilities

This section, originally enacted in 1984, includes a number of relevant provisions. It

1. discusses the necessity for core, government-owned and -operated logistics capabilities (employing government personnel and equipment)
2. directs the Secretary of Defense to identify core logistics capabilities
3. defines core logistics capabilities as those necessary to maintain and repair weapon systems and other military equipment (including mission-essential weapon systems or materiel, no later than four years after achieving IOC, but excluding systems and equipment under special access programs, nuclear aircraft carriers, and certain commercial items)
4. requires the secretary to ensure that the core logistics workloads necessary to maintain core logistics capabilities are performed at government-owned and -operated DoD facilities of DoD (including those belonging to a military department)
5. requires the secretary to assign such facilities sufficient workload to ensure cost efficiency and technical competence in peacetime while preserving the surge capacity and reconstitution capabilities necessary to support strategic and contingency plans
6. precludes this workload from being competed with nongovernment personnel under Office of Management and Budget (OMB) Circular A-76 procedures
7. gives the secretary waiver authority and procedures for implementing it for certain workloads not required for national defense reasons
8. contains restrictions on DoD entering into a prime vendor contract for depot-level maintenance and repair.

10 USC 2466, Limitations on the Performance of Depot-Level Maintenance of Materiel

This section discusses limitations on the amount of depot-level maintenance and repair workload that contractors, as opposed to government facilities, can perform. The current limit is 50 percent of the funds for depot-level maintenance and repair workload per military department or defense agency. This workload restriction was originally established in 1988. The Secretary of Defense is allowed to waive this limitation for a fiscal year if he or she determines that the waiver is necessary for reasons of national security and if he or she submits to Congress a notification of the waiver together with the reasons for it. This section also requires an annual report that identifies the total amount expended for depot-level maintenance and repair, as well as how much is spent or is planned to be spent on public as opposed to private-sector activities in the prior, current, and ensuing fiscal years. In addition, it requires the Comptroller General to complete a review of this report within 90 days of its submission.

10 USC 2469, Contracts to Perform Workloads Previously Performed by Depot-Level Activities of the Department of Defense: Requirement of Competition

This section requires the Secretary of Defense to ensure that depot-level maintenance and repair workload is not transferred to a contractor or another depot-level DoD activity unless the change is made using (1) merit-based selection procedures for competitions among all DoD depot-level activities or (2) procedures for competitions among private and public-sector entities. This restriction applies to any workload greater than $3 million that is being performed by a DoD activity. A waiver provision addresses public-private depot partnerships.

10 USC 2470, Depot-Level Activities of the Department of Defense: Authority to Compete for Maintenance and Repair Workloads of Other Federal Agencies

This section, enacted in 1994, allows DoD depot-level activities to compete for the performance of any depot-level maintenance and repair workload of a federal agency that uses competitive procedures to select the performer.

10 USC 2472, Prohibition on Management of Depot Employees by End Strength

This section mandates that civilian employees of DoD who perform, or are involved in the performance of, depot-level maintenance and repair workloads must be managed solely on the basis of the available workload and the funds available for depot-level maintenance and repair. These government employees cannot be managed on the basis of any constraint or limitation in terms of man years, end strength, full-time equivalent positions, or maximum number of employees.

10 USC 2474, Centers of Industrial and Technical Excellence: Designation; Public-Private Partnerships

This section directs the Secretary of Defense to designate each DoD depot-level activity (other than facilities approved for closure or major realignment under the Defense Base Closure and Realignment Act of 1990) as a Center of Industrial and Technical Excellence in its recognized core competencies. It also directs the secretary to establish a policy to encourage each military department and defense agency to reengineer industrial processes and adopt best business practices at its Centers of Industrial and Technical Excellence.

10 USC 2474 allows the military departments to conduct pilot programs to test any practices that could improve the efficiency and effectiveness of operations at the Centers of Industrial and Technical Excellence, improve the support these centers provide, and enhance readiness by reducing the time it takes to repair equipment.

The section authorizes the head of each center to enter into public-private cooperative arrangements to conduct depot-level maintenance and repair activities related to its core competencies and establishes procedures for doing this. The amounts expended for nongovernment employees during fiscal years 2003–2009 do not count for 50-50 law compliance purposes if the personnel are provided by private industry or other entities outside DoD pursuant to a public-private partnership. These amounts are reported as a separate item in the annual report to Congress.

10 USC 2563, Articles and Services of Industrial Facilities: Sale to Persons Outside the Department of Defense
Under special conditions, this statute allows a working capital–funded industrial facility to sell articles that are not available commercially in the United States to a purchaser other than DoD.

Government Accountability Office Reports

Given the congressional interest in DoD depot-level maintenance and repair activities, GAO has written many reports over the last 30-plus years on the subject. The office has addressed four major recurring themes in the area of Air Force depot maintenance:

1. the inaccuracy of the accounting and reporting systems, particularly in light of the requirements of the Federal Manager's Financial Integrity Act of 1982 and the Chief Financial Officer's Act of 1990
2. the overuse of ICS for new weapon systems
3. insufficient use of competition in determining the proper source of repair
4. interservice normalization of workload and elimination of excess capacity.

This section of the appendix provides a brief synopsis of the most relevant GAO reports.

Should Aircraft Depot Maintenance Be In-House or Contracted? Controls and Revised Criteria Needed (GAO, 1976)
GAO (1976) notes that DoD had two sets of policies on distributing depot maintenance workloads between organic depots and contractors. The first, found in DoDD 4100.15 and DoDI 4100.33, was based on OMB Circular A-76 guidance, which called for use of private enterprise in satisfying military needs except when organic support was needed for combat support, retraining military personnel, or retention or strengthening of mobilization readiness or when procurement from private enterprise would have increased costs. The other set of policies

(found in DoDD 4151.1) directed that organic capacity be planned to accomplish a maximum of 70 percent of the gross mission essential workload.

GAO believed this DoD guidance was unclear because it implied that all non–mission-essential workload and at least 30 percent of the mission-essential workload should be performed at commercial sources. GAO found that a far smaller share of the workload was being performed commercially. It also noted that military departments distributed workload by filling organic capacity first, then contracting the remainder. This followed neither A-76 guidance, which emphasized comparative costs, nor DoD guidance on the 70-30 split. The office found that the Air Force was within the 70-30 guidance but that a considerable portion of the organic workload was non–mission-essential. The services made few cost comparisons when distributing workload between organic and commercial sources. GAO recommended that the 70-30 policy be reconsidered and that criteria be developed to

- assess time-phased mobilization surge needs at organic depots and develop a goal for a minimum organic and contractor capacity to meet the needs, as well as to relate this capacity to peacetime workloads
- determine what types of materials should be supported organically
- determine when cost versus mission essentiality for distributing workload should apply
- require the military departments to apply controls for following DoD policies on planning organic capacity and distributing workload.

Statement Before the Subcommittee on Legislation and National Security of the House Government Operations Committee (Gilroy, 1984)

Robert Gilroy told the committee that the amount DoD spent on commercial maintenance, repair, and modification had increased over the preceding few years, partially due to the Executive Branch's policy of increasing commercial contracting and partially due to manpower ceilings and other DoD resource limitations.

The statement addressed the same issue as GAO (1976), that DoD had two sets of policy guidance: one that applied to all DoD commercial and industrial activities (the OMB Circular A-76 series) and the other that applied specifically to DoD maintenance activities. The relationship between the two had not been clear historically. Gilroy noted that OMB Circular A-76 specified the government policy for obtaining goods and services from the private enterprise system, with exceptions when no commercial source was available or if required for national defense reasons. He also noted that the then–most recent version of the circular (August 1983) had dropped the previous exemption for military intermediate- and depot-level maintenance for national defense reasons. DoD had not issued new criteria for determining when government performance was needed for national defense reasons. Circular A-76 also allowed an exception from commercial sourcing when it could be demonstrated that government performance was less costly.

The other guidance from DoD (in DoDD 4151.18) required planning depot maintenance for both contractual services and in-house work, but did not specify how to determine the mix. It specified that contractor maintenance should be cost-effective but did not specify how cost-effectiveness should be determined. GAO found DoD did not evaluate the relative costs of contract and in-house maintenance consistently and, in some cases, had awarded the work to a higher-cost provider. GAO also found that DoD had not followed the OMB Circular A-76 instructions, especially its definition of a *commercial activity*. GAO noted that DoD had no consistent way of determining the amount of in-house work required for flexibility and rapid surge.

Gilroy stated that DoD was not competing enough of the contract services workload among potential sources and that it was using contracting vehicles that are difficult to administer, such as basic ordering agreements and time and materials contracts. In FY 1982, 65 percent of the $4.8 billion spent on contractor repair services had been awarded noncompetitively. GAO also noted that many of the competitively awarded contracts had only one bidder. Roughly 55 percent of the $4.8 billion total (or $2.7 billion) was spent on weapon system–

related services. Of that total, 86 percent was awarded noncompetitively. The Air Force figure was 74 percent for noncompetitive awards, with significant variance among the ALCs.

Gilroy also noted that purchases of spare parts were historically noncompetitive, as the GAO, the DoD/IG, and the Air Force Audit Agency had found in previous reports. One major reason was a lack of sufficient technical data to allow spare parts procurements to be competed. This same lack of technical data was affecting competitions for maintenance workloads. Gilroy stated that, when costs increased during the acquisition phase of a weapon system procurement, funds for establishing in-house capabilities were often decreased, thereby delaying that capability and therefore the ability to compete the workload.

As an example of the inadvisability of using time-and-materials contracts, Gilroy cited the F-15 Pacer Webb contract. Awarded in 1975 as a temporary contract, it had been extended for more than seven years because of a lack of in-house capability. The Air Force did not have contractor unit-repair cost data because it was a time-and-materials contract, so it was not possible to determine whether the payments were reasonable. In addition, time-and-materials contracts provided disincentives for contractors to reduce costs.

Potential for Improving Depot Maintenance Productivity Measurement and Reporting (Conahan, 1985)

In his letter to the Assistant Secretary of Defense for Manpower, Installations, and Logistics, Frank Conahan stated GAO's opinion that the Defense Productivity Program Office's measurement system data were unusable and that the Depot Maintenance and Maintenance Support Cost Accounting and Production Reporting System specified in DoD 7220.29-H would be a better source of consistent, complete, and reliable productivity data. GAO recommended development of proper formats to be used to satisfy the Bureau of Labor Standards productivity reporting requirement and for OSD management of organic depot maintenance activities.

Strategic Bombers: B-1B Maintenance Problems Impede Its Operations (GAO, 1988)

GAO noted that B-1B maintenance costs were increasing, that reliance on contractor engineering support would be extended, and that there were significant maintainability challenges. Problems with the onboard test systems were affecting maintainability. The Air Force had not received the planned support and maintenance instructions to allow organic maintenance, which slipped the transition schedule two years. As a result, ICS costs had increased from the estimated $250 million to $570 million, with the only organic depot maintenance being for the B-1's engines. All other depot maintenance was being performed under ICS. In addition, sustaining engineering costs had increased by about 140 percent above the estimated cost. Poor mission-capable rates were affecting aircraft availability for alert and aircrew training.

Financial Audit: Air Force Does Not Effectively Account for Billions of Dollars of Resources (GAO, 1990a)

This audit of the 1988 Air Force financial statements found inaccuracies and incorrect figures for most noncash assets. With poor cost data, it was difficult to make the kind of cost comparisons OMB Circular A-76 requires for determining whether contractor or organic depots should perform repairs. GAO noted that the "[o]perating costs of air wings, depots, and commands cannot be compared and evaluated." (GAO, 1990a, p. 8). Many of the same findings of this audit were repeated in a subsequent audit (GAO, 1992a).

Military Bases: Information on Air Logistics Centers (GAO, 1990b)

This report summarizes performance and capacity data on the five Air Force ALCs from FY 1985 through FY 1989. It contained only organic workloads, not contractor or interservice workloads. Some of the report's observations were that the average daily output of direct labor per maintenance employee was about four hours at all ALCs; the average age of the facilities was about 30 years; the average age of the maintenance equipment was about 12 years; and the ALCs' expenses were $169 million higher than their revenues in FY 1989.

Contract Maintenance: Improvement Needed in Air Force Management of Interim Contractor Support (GAO, 1992b)

GAO noted that Air Force ICS had grown dramatically since its 1983 report. Between 1985 and 1992, the cost of Air Force ICS had tripled (to $328 million), and the number of systems under ICS had quintupled (to 48). The B-1, B-2, and F-15 had the most ICS support by dollar value in FY 1992. The C-17 program, although successfully transitioning to organic capability at operating bases, was not as successful in transitioning to depot maintenance. GAO found the same problems behind excessive ICS activities as it had in its 1983 report. An Air Force ICS study team developed a couple of initiatives to address this, including funding ICS from the procurement accounts rather than the O&M account (implemented in the FY 1993 appropriation) and including depot support requirements in operational requirements documents. The report noted that funding of ICS in procurement accounts would make it more difficult to track total ICS and total maintenance costs.

Air Logistics Center Indicators (GAO, 1993)

GAO was asked to obtain information on workload, productivity, quality, capacity, and financial indicators at Air Force ALCs. GAO noted the lack of an information system and standardized procedures that could provide consistent, comparable data in these areas. In 1990, DoD had initiated the Depot Maintenance Performance Measurement System, but the system was not yet in place at the time GAO (1993) was being written. GAO noted that, despite efforts since the 1960s, depot maintenance capacity still exceeded requirements by an estimated 25 to 50 percent of foreseeable workload. The Air Force had chosen to reduce facilities at its ALCs rather than closing bases. Relative to a baseline workload in 1987, the Air Force workload was 21 percent lower in 1992, with a projection of 32 percent by 1997. The GAO also reiterated that the cost data available to depot management was inaccurate.

Depot Maintenance: Issues in Management and Restructuring to Support a Downsized Military (Heivilin, 1993)

Donna Heivilin reported that GAO found, among other things and most pertinent to this monograph, DoD did not have a comprehensive strategy for determining what depot maintenance workload should be performed by the private sector and that public-private competitions had not been implemented consistently across the services.

Heivilin noted that ICS had been a common practice, but for some systems, such as the B-1B, it continued for far too many years.

The Defense Depot Maintenance Council Corporate Business Plan defined core requirements, but despite DoD guidance, the services had not made core workload determinations. Legislative restrictions at that time were that only 40 percent could be awarded to the private sector. 10 USC 2464 required DoD activities to maintain a logistics capability sufficient to ensure technical competence and the resources necessary for a response to a mobilization or emergency. Although the Secretary of Defense was required to identify these activities, this had not been done.

DoD Directive 4151.01, *Use of Contractor and DoD Resources for Maintenance of Materiel,* issued in 1982, continued the requirement that in-house work should be kept to the minimum necessary to meet military contingencies. It also stated that, to the extent possible, a competitive commercial depot maintenance industrial base should be established. More specifically, it provided that prime consideration should be given to use of contractor support when it would improve the industrial base, improve peacetime readiness and combat sustainability, be cost-effective, or promote contract incentives for reliability and maintainability. To some extent, this directive also retained the previously established 70-30 ratio (Heivilin, 1993, p. 14).

This DoD directive was effectively superseded by a 1992 amendment to 10 USC 2466 that prohibited the military departments from contracting out more than 40 percent of their respective depot maintenance work for performance by the private sector. Section 2466 provided that the service secretaries and the Secretary of Defense could waive this restriction if the Secretary determines the waiver is necessary for national security reasons and notified the Congress of the reasons for the waiver (Heivilin, 1993, p. 16).

Heivilin noted that the public-private competitions had achieved very little actual savings but also that the projections of savings from implementing additional competitions were overly optimistic. Private industry was dubious about the likelihood of a "level playing field" in the decisions about workload.[1] She noted that, of all the DoD competitions held, the private sector had won about 60 percent (although the private sector had won only 38 percent of competitions in the Air Force). But of all these competitions, only nine workloads had shifted from one sector to the other, with eight moving from public to private (Heivilin, 1993, p. 23).

Depot Maintenance: Issues in Allocating Workload Between the Public and Private Sectors (Heivilin, 1994)

This testimony addressed questions about how much workload should be retained in public depots as core capability, whether a service should be allowed to have its own core capability, and how the remaining noncore workload should be allocated between the public and private sectors. Although DoD had reported a 65-35 public-private workload allocation, GAO estimated it was 50-50. Heivilin (1994) noted that DoD had 35 organic depots at the beginning of the BRAC process; by the completion of the projected BRAC closures, the number would be 24. But there was still excess capacity.

Heivilin noted that a portion of the funds expended on the organic workload were ultimately contracted out to the private sector for parts and material, maintenance and engineering services, and other goods and services. However, these funds were included in the public sector's share of depot maintenance expenditures. In addition, some types of depot maintenance activities, such as ICS, were not included in the statistics. She also noted inconsistencies in how the services collected and aggregated data to develop DoD's report to Congress on the public and private mix for depot-level maintenance. Based on GAO's calcula-

---

[1] Heivilin, 1993, pp. 17–19, provides a history of the impetus behind these competitions.

tions, only 43 percent of the Air Force's funding in FY 1993 went to the public sector, with the remainder going to private firms.

GAO recommended each service develop core requirements but did not support the concept that each service should necessarily have core capacities as long as they existed somewhere in DoD's depots.

Depot Maintenance: Management Attention Required to Further Improve Workload Allocation Data (GAO, 2001b)

10 USC 2466 contains the 50-percent limitation on private-sector depot maintenance and requires DoD to submit two reports on public and private-sector depot maintenance workloads to Congress every year. The first report provides the percentages of funds expended in the public and private sectors during the two preceding fiscal years, and the second report projects the same information for the current and four succeeding fiscal years.

The section also requires GAO to give Congress its views on whether DoD complied with the 50-50 requirement in the prior-years report and whether the projections in the future-years report were reasonable. GAO (2001b) therefore discussed whether (1) the military departments met the 50-50 requirement for FYs 1999 and 2000 and (2) the projections for FYs 2001 through 2005 represented reasonable estimates.

As a part of this work, GAO also examined DoD's efforts to improve the reporting process and sought to identify opportunities to further improve it. GAO analyzed each service's procedures and internal management controls for collecting, aggregating, and reporting depot maintenance information for responding to the section 2466 requirements. GAO reviewed its previous year's report covering the FY 1998 and 1999 prior-year workloads and FY 2000–2004 future-year workloads and noted that, because of the limitations of the DoD depot maintenance reporting data, its analysts were unable to determine precisely whether DoD had complied with the 50-percent limitation. The report recognized the limitations of DoD's financial systems and data and noted that, in addition to the data reliability weaknesses, their audits of financial management operations routinely identified pervasive weaknesses in financial systems and fund controls that adversely affected DoD's ability to accumulate costs and reliably determine expenditures, obligations, and funding availability.

GAO's analysis of DoD's 50-50 data showed that the quality of the data reported to Congress in 2000 had substantially improved over previous years, but the analysts continued to find errors and inconsistencies in the reporting and in how well the services documented their analyses supporting their workload reports. GAO also recommended that the Air Force implement a long-term strategy to comply with the 50-percent requirement.[2]

Depot Maintenance: Change in Reporting Practices and Requirements Could Enhance Congressional Oversight (GAO, 2002)

This document is similar to GAO reports from previous years on DoD's reporting of its compliance with 10 USC 2466.

Depot Maintenance: Persistent Deficiencies Limit Accuracy and Usefulness of DoD's Funding Allocation Data Reported to Congress (GAO, 2005)

As in previous years, systemic weaknesses in DoD's financial systems and persistent deficiencies in 50-50 data reporting processes continued to prevent GAO from determining whether the military departments had complied with the 50-50 requirement for public and private-sector depot maintenance funding allocations for FY 2004. GAO's reports over the previous seven years had identified similar problems and recommended corrective actions, but DoD and the military services failed to implement corrective actions consistently that were sufficient to resolve the deficiencies and alleviate data accuracy problems. That these problems were recurring indicated a management control weakness as defined under the Federal Managers' Financial Integrity Act.

---

[2] For a similar analysis, see GAO, 2000a, and GAO, 2000b.

OSD Guidance

To implement congressional direction and to reflect the management direction of the Secretary of Defense, OSD has promulgated a number of directives, instructions, and other forms of guidance to the services concerning the functioning of depot-level maintenance and repair. The following is a summary, in type and numerical order, of the major OSD policy documents most relevant to this report.

DoDD 4151.18, Maintenance of Military Materiel (2004)

This directive promulgates the requirements of 10 USC 2464 for inherently governmental and core-capability requirements and for noncore capability requirements under competitive sources in accordance with 10 USC 2462 and 10 USC 2466. It directs that initial maintenance program management begin at the initiation of program acquisition activities, with core depot capability requirements to be identified as early as possible in the acquisition life cycle. The directive also requires establishing core capabilities no later than four years after IOC. It requires the individual DoD components to identify core capabilities and calculate depot workloads associated with them and to designate a major organic depot activity as a Center of Industrial and Technical Excellence in one or more specific technical competencies required for core capabilities.

DoDD 4275.5, Acquisition and Management of Industrial Resources (2005)

This directive provides policy guidance on acquiring and managing facilities, special tooling, and special test equipment, whether used by organic activities or by contractors.

DoDD 5000.1, The Defense Acquisition System (2003)

This directive is the bible of the defense acquisition system and establishes overall policy on how the system should operate. It addresses the requirements for the development, production, and sustainment phases of each acquisition program. It directs use of PBL to optimize total system availability while minimizing cost and logistics footprints.

DoDD 5128.32, Defense Depot Maintenance Council (1990)

This directive establishes the Defense Depot Maintenance Council, with the missions of advising the Deputy Under Secretary for Logistics on depot maintenance initiatives for reducing costs and improving efficiency of management, serving as a review mechanism for policies, and serving as an exchange forum for operations.

DoD Handbook 4151.18-H, Depot Maintenance Capacity and Utilization Measurement Handbook (1997)

This handbook provides specific guidance for measurement of the capacity and utilization of organic depot maintenance activities.

DoDI 4000.19, Interservice, Interdepartmental and Interagency Support (1995)

This instruction addresses procedures for obtaining or providing logistics and other support from other services or other government departments, reimbursement policies for incremental direct costs incurred by the provider, and other provisions.

DoDI 4100.33, Commercial Activities Program Procedures (1985)

This instruction describes the policies, procedures, and responsibilities DoD uses to determine whether commercial activities should be performed organically or commercially. The directive states that functions that are "inherently governmental in nature, and intimately related to the public interest," must be performed by "DoD personnel only" (p. 3). It allows organic commercial activities to remain in-house when the activity is essential for training or experience in required military skills, needed to provide appropriate work assignments for a rotation base for overseas or sea-to-shore assignments, or necessary to provide career progression to needed military skill levels.

On p. 3, the instruction states that

> DoD Components shall rely on commercially available sources to provide commercial products and services, except when required for national defense, when no satisfactory commercial source is available, or when in the best interest of direct patient care. DoD Components shall not consider an in-house new requirement, an

expansion of an in-house requirement, conversion to in-house, or otherwise carry on any CAs [commercial activities] to provide commercial products or services if the products or services can be procured more economically from commercial sources.

Finally, this instruction lays out the requirements for the cost comparisons that must be performed for noncore workload, including congressional notification if the workload being studied for conversion to contract involves 46 or more DoD civilian workers.

DoDI 4151.20, Depot Maintenance Core Capabilities Determination Process (2007)

This instruction assigns responsibilities in OSD and the components for determining and reporting core depot maintenance capabilities.

DoDI 5000.2, Operation of the Defense Acquisition System (2003)

This instruction describes the DoD decision process for system acquisition, milestone requirements, system sustainment requirements, and sustainment strategies. It advocates evolutionary acquisition. It does not specifically mention CLS or depot maintenance requirements.

DoDI 5154.19, Defense Logistics Studies Information Exchange (DLSIE) (1972)

This instruction specifies how any studies, logistics research, and management information must be reported, stored, and distributed under the Department of the Army, which is the executive agent for DoD.

DoD Regulation 4140.1-R, DoD Supply Chain Material Management Regulation (2003)

This regulation provides guidance for developing material requirements, selecting support providers based on best value, determining how to position and deliver material, and executing supply chain functions. It allows material managers to establish commercial support agreements or partnerships but does not address selection procedures or 50-50 constraints.

DoD Regulation 7000.14-R, Department of Defense Financial Management Regulations (various dates)

This regulation provides guidance on working-capital procedures, budgets, and payment procedures.

DoD Inspector General Reports on Air Force CLS

The DoD IG is the investigative arm of OSD and regularly investigates situations and activities within the services and DoD agencies. The following two recent reports relate to CLS in the Air Force and are included to illustrate the difficulty of demonstrating the performance or cost benefits of CLS.

Implementation of Performance-Based Logistics for the Joint Surveillance Target Attack Radar System (DoD IG, 2006c)

The Air Force stated that it had implemented PBL on the system in June 2004. The IG found that the program manager had not fully implemented PBL and could not show where reduced life-cycle costs or increased system availability had occurred.

Procurement Procedures Used for C-17 Globemaster III Sustainment Partnership Total System Support, D-2006-101 (DoD IG, 2006b)

The IG criticized the Air Force for awarding a total system support responsibility contract to Boeing for FYs 2004–2011 without conducting a business-case analysis or considering alternatives to a sole-source contract with the OEM. Neither did the Air Force consider core and noncore workloads in the analysis.

Air Force Guidance

With congressional direction and OSD guidance, the Air Force has published a number of instructions and regulations that address depot maintenance requirements and capabilities, as well as processes for making decisions about sources of these capabilities. The following is a list of the major regulations and a summary of how each affects depot maintenance.

AFI 10-602, Determining Mission Capability and Supportability Requirements (2005)

This instruction provides procedures and parameters that define and maximize the capability and supportability of the logistics mission throughout a system's life cycle. It directs the "single manager" to develop a SORAP recommendation, which is to be briefed to the Acquisition Strategy Panel. It requires a product support management plan. The instruction further directs development of a "best value" depot maintenance decision, including a "core analysis," a 50-50 assessment, and a review of organic and contract capabilities.

AFI 20-104, System Executive Management Report (1998)

This instruction addresses semiannual sustainment and readiness reporting for Air Force weapon systems. The term *contractor logistics support* is shown in Attachment 2 as one of three contractor services, along with ICS and miscellaneous contract services.

AFI 21-102, Depot Maintenance Management (1994)

This instruction assigns responsibilities for depot maintenance management to HQ USAF Logistics Maintenance Management, HQ USAF Logistics Plans (Logistics Readiness Center), and HQ AFMC. It assigns responsibility to HQ AFMC to develop and maintain

> a methodology for assessing organic depot maintenance minimum level requirements and making depot maintenance source of repair (SOR) determinations in accordance with criteria established by DoD Directive 4151.18. . . . .

It also requires HQ AFMC, in conjunction with each ALC, to establish a comprehensive depot maintenance program for all new system acquisitions, including logistics management for the life of the system, interim support arrangements, and the ultimate maintenance concept.

Chapter 2, "Business Planning," addresses the depot maintenance activation planning as a new weapon system progresses through development, how the HQ AFMC Business Board decides on the source of repair for depot workloads, and the responsibilities and tasks required for activating depot support for a new system. It directs that, when source-of-repair decisions call for an organic depot, the capability must be established no later than IOC.

Chapter 3, "Sources of Repair," discusses the philosophy for assignment of workloads between organic and private-sector contractors; requires an ALC as a responsible agent for the entire depot maintenance workload, regardless of the source of repair; notes that ALCs accomplish high, surge workloads organically as a general rule and that ALCs can also perform complimentary high-volume peacetime workloads for which the technology or skills are similar but that do not have a high wartime requirement. The regulation states that organic depot maintenance workload should be sized to 100 percent of the workload for one 40 hour-per-week shift and that the shop's planned maximum utilization limit should be 250 percent of physical capacity. AFI 21-102 (1994, p. 7), notes that

> CLS performs many functions normally accomplished by an organic support activity, including item management, supply, distribution, repair, depot maintenance, operating command organizational and intermediate levels of maintenance as negotiated, and many other operations and maintenance tasks. CLS principally supports depot field teams, low surge workloads, small workloads, commercial off-the-shelf items, and short life cycle or rapid obsolescence items. Consider use of CLS for high-surge workloads that either involve unique processes, for capabilities that cannot be established organically at reasonable cost or other factors that clearly establish CLS to be in the best interest of the Air Force by virtue of lower costs and/or increased readiness.

AFI 21-102 (1994, p. 16) defines CLS as "A preplanned contractor support method used to provide all or part of the ILS elements for a system, equipment, or item for long periods of time or until retirement."

The instruction states that POS or ICS for depot maintenance may be used prior to an organic capability being ready.

Attachment 2 to AFI 21-102 (1994) describes a decision-tree process for selecting either organic or contract depot maintenance as a source of repair, which is illustrated in the instruction's Fig. A2.1.

AFMC Instruction 21-101, Depot Maintenance Activation Planning (1994)

The logistics support analysis process determines the planning requirements for depot maintenance activation. This instruction states that a depot maintenance activation plan will be developed for each depot activation and describes organizational responsibilities in the process.

AFI 63-107, Integrated Product Support Planning and Assessment (2004)

This instruction places responsibility for both acquisition and sustainment product-support planning on the program manager. It addresses the LCMP, which integrates both the acquisition and sustainment strategies from program concept development to disposal, and provides all support requirements for a weapon system, subsystem, or major item and is part of the exit criteria from Milestone B and later. It also states that PBL is the preferred approach for implementing product support and must be used on all new acquisition category I and II systems. It presents the mission-assignment process, in which a system is assigned to a specific ALC for life-cycle support prior to Milestone B. AFI 63-107 (2004) Chapter 4 discusses public-private partnerships and their use as a means to obtain best value depot maintenance for noncore workloads. The SORAP is discussed at length in Chapter 5. Chapter 6 addresses the requirements of migration planning, in which a weapon system is retired and stored.

AFI 63-101, Operations of Capabilities Based Acquisition System (2005)

This instruction promulgates the direction for conducting systems acquisitions specified by AFPD 63-1, "Capabilities-Based Acquisition System," and the DoD 5000 series. It stipulates that logistics elements be considered and included during all phases of acquisition planning and that PBL is the preferred strategy for delivering sustainment resources at minimum life-cycle cost. The LCMP (required for acquisition category I and II programs) is the means of integrating acquisition and sustainment strategies and requires an update prior to each milestone decision. The instruction assigns roles and responsibilities in the acquisition process. The program manager is tasked with "fully identifying" all Program Support Concept elements at Milestone B. The program manager is to develop and document an acquisition and sustainment strategy that is approved by the milestone decision authority at each milestone review. After Milestone C, program management is transferred from the PEO to an ALC sustainment management portfolio when the system is deemed ready. CLS is not specifically mentioned in the instruction.

AFI 63-111, Contractor Support for Systems, Equipment and End Items (2005)

This instruction implements the requirements of AFPD 20-5 and AFPD 63-1. It provides policies and procedures for funding, implementing, and managing contractor support throughout the life cycle of systems, equipment, and end items or for their modification or upgrade. It defines three types of contract support: POS, ICS, and CSS. POS is used to support test and evaluation efforts, risk reduction efforts, production readiness, and other temporary periods during the acquisition or modification of the system and is addressed during Milestone A and defined prior to Milestone B. ICS is temporary support for an initial period of the operation of the system. The ICS strategy must include a plan to transition to either organic or contract sustainment (or a combination of both). CSS is used when organic life-cycle logistics support is not planned. Decisions on organic or contract support can be revisited later.

CSS requirements are identified based on the type of funding used. When the funding is for multiple sustainment elements, the source of funds is O&M appropriations using Air Force Element of Expense (AFEE) 578, CLS; for this element, the system program manager will document the defined support annually in a requirements brochure (CLS brochure), which addresses nine years of requirements. In some cases, multiple CSS is accounted for under AFEE 592, Miscellaneous

Contract Services funds. When support is for a single sustainment element, the source of funds is usually for the specific element, such as AFEE 583, Sustaining Engineering By Contract; AFEE 594, Contract Technical Data; or AFEE 560 or 54x, Depot Maintenance Provided Through the Depot Purchased Equipment Maintenance Program.

AFPD 20-5, Air Force Product Support Planning and Management (2001)

This directive establishes the framework for implementing product-support management in the Air Force in response to Section 912 (c) of the National Defense Authorization Act for FY 1998, which required DoD to submit a plan to Congress for streamlining acquisition organizations, workforces, and infrastructure. The directive encompasses the instructions provided in AFI 63-107 for integrated product-support planning, AFI 63-111 for contractor support planning, and AFI 21-102 for depot maintenance management.

The directive requires that product support begin early in the acquisition phase of a weapon system, preferably in the concept and technology development phase, and that the transition to sustainment be seamless. It defines *product support* as "the package of support functions necessary to maintain the readiness and operational capability of weapon systems, subsystems, and support systems." The source of support may be organic or commercial, but its primary focus is on optimizing customer support and achieving "maximum weapon system availability at the lowest total ownership cost."

AFDD 20-5 requires single managers to document the strategy in a product-support management plan, which is considered a living document that should be updated as the weapon system transitions through its successive phases.

Finally, the directive defines CLS as a "planned cost effective contract support method used to provide all or part of the logistics support elements for a system, equipment, or item for extended periods of time or for the life cycle of the system or equipment."

AFPD 21-1, Air and Space Maintenance (2003)

This directive establishes policy and assigns responsibilities for the maintenance of air and space equipment to meet operational needs, including mobilization and surge requirements. It requires maintaining a depot maintenance capability to meet military contingency requirements and using performance-based agreements, including partnerships, to achieve economy and efficiency. In addition, it directs the establishment of inter- and intraservice and joint contracting maintenance support agreements. It directs AFMC to manage depot maintenance activities and to determine core capabilities annually.

References

AFI—*See* Air Force Instruction.

AFPD— *See* Air Force Policy Directive.

AFTOC OLAP—*See* Air Force Total Ownership.

Air Force Cost Analysis Agency, Force Analysis Division, Factors Branch (FMFF), Air Force Instruction 65-503 Table A7-1, *Contractor Logistics Support (FY08)*, February 2007.

Air Force Logistics Management Agency, *Maintenance Metrics US Air Force*, December 20, 2001.

Air Force Total Ownership Cost, Online Analytic Processing, management information system, January 8 and May 2007.

Air Force Instruction 10-602, *Determining Mission Capability and Supportability Requirements*, Washington, D.C.: Department of the Air Force, March 18, 2005.

————, 20-104, *System Executive Management Report*, Washington, D.C.: Department of the Air Force, December 11, 1998.

————, 21-102, *Depot Maintenance Management*, Washington, D.C.: Department of the Air Force, July 19, 1994.

————, 21-133(I), *Joint Depot Maintenance Program*, Washington, D.C.: Department of the Air Force, March 31, 1999.

————, 63-101, *Operations of Capabilities Based Acquisition System*, Washington, D.C.: Department of the Air Force, July 29, 2005.

————, 63-101, *Acquisition and Sustainment Life Cycle Management*, draft, Washington, D.C.: Department of the Air Force, May 27, 2008.

————, 63-107, *Integrated Product Support Planning and Assessment*, Washington, D.C.: Department of the Air Force, November 10, 2004.

————, 63-111, *Contractor Support for Systems, Equipment and End Items*, Washington, D.C.: Department of the Air Force, October 21, 2005.

Air Force Materiel Command, *Contractor Logistics Support (CLS) Requirements Handbook*, initial draft, May 2008.

Air Force Materiel Command Instruction 21-101, *Depot Maintenance Activation Planning*, February 18, 1994.

Air Force Policy Directive 20-5, *Air Force Product Support Planning and Management*, Washington, D.C.: Department of the Air Force, April 10, 2001.

————, 21-1, *Air and Space Maintenance*, Washington, D.C.: Department of the Air Force, February 25, 2003.

————, 63-1, *Capabilities-Based Acquisition System*, Washington, D.C.: Department of the Air Force, July 10, 2003.

Assistant Secretary of the Air Force, Acquisition, "Flexibility in Contractor Logistics Support (CLS) Contracts," memorandum, Washington, D.C.: Department of the Air Force, June 4, 2008.

Coale, Scott G., and George Guerra, "Transitioning an ACTD to an Acquisition Program: Lessons Learned from Global Hawk," *Defense AT&L*, September–October, 2006, pp. 7–11.

Conahan, Frank C., "Potential for Improving Depot Maintenance Productivity Measurement and Reporting," letter to the Assistant Secretary of Defense for Manpower Installations and Logistics, Washington, D.C.: General Accounting Office, GAO/GGD-85-49, April 1985.

Cook, Cynthia R., John A. Ausink, and Charles Robert Roll, Jr., *Rethinking How the Air Force Views Sustainment Surge*, Santa Monica, Calif.: RAND Corporation, MG-372-AF, 2005. As of September 19, 2008: http://www.rand.org/pubs/monographs/MG372/

Department of the Air Force, Contractor Logistics Support Brochures, various programs, FYs 2006, 2007, and 2008.

Department of Defense, *Quadrennial Defense Review Report*, Washington, D.C., September 30, 2001.

Department of Defense Directive 4151.18, *Maintenance of Military Materiel*, Washington, D.C., March 31, 2004.

————, 4275.5, *Acquisition and Management of Industrial Resources*, Washington, D.C., March 15, 2005.

————, 5000.1, *The Defense Acquisition System*, Washington, D.C., May 12, 2003.

————, 5128.32, *Defense Depot Maintenance Council*, Washington, D.C., November 7, 1990.

Department of Defense Handbook 4220.29-H, *Depot Maintenance Support Cost Accounting and Production Reporting Handbook,* Washington, D.C., October 1, 1975.

————, 4151.18-H, *Depot Maintenance Capacity and Utilization Measurement Handbook,* Washington, D.C., January 1997.

Department of Defense Instruction 4000.19, *Interservice, Interdepartmental and Interagency Support,* Washington, D.C., August 9, 1995.

————, 4100.33, *Commercial Activities Program Procedures,* Washington, D.C., September 9, 1985.

————, 4151.20, *Depot Maintenance Core Capabilities Determination Process,* Washington, D.C., January 5, 2007.

————, 5000.2, *Operation of the Defense Acquisition System,* Washington, D.C., May 12, 2003.

————, 5154.19, *Defense Logistics Studies Information Exchange (DLSIE),* Washington, D.C., July 13, 1972.

Department of Defense Regulation 4140.1-R, *DoD Supply Chain Material Management Regulation,* Washington, D.C., May 23, 2003.

————, 7000.14-R, *Department of Defense Financial Management Regulations,* Washington, D.C., various dates.

Department of Defense, Office of the Inspector General, *The Military Departments' Implementation of Performance-Based Logistics in Support of Weapons Systems,* Washington, D.C., D-2004-110, August 23, 2004.

————, *Air Force Procurement of 60K Tunner Cargo Loader Contractor Logistics Support,* D-2006-059, Washington, D.C., March 3, 2006a.

————, *Procurement Procedures Used for C-17 Globemaster III Sustainment Partnership Total System Support,* Washington, D.C., D-2006-101, July 21, 2006b.

————, *Implementation of Performance-Based Logistics for the Joint Surveillance Target Attack Radar System,* Washington, D.C., D-2006-105, August 9, 2006c.

Deputy Assistant Secretary of the Air Force, Contracting, "Acquisition of Technical Data for Commercial and Commercial Derivative Aircraft," *memorandum,* Washington, D.C.: Department of the Air Force, February 11, 2002.

DoDD—*See* Department of Defense Directive.

DoDI—*See* Department of Defense Instruction.

DoDIG—*See* Department of Defense, Office of the Inspector General.

Drezner, Jeffrey A., and Robert S. Leonard, *Innovative Development: Global Hawk and Dark Star—Transitions Within and Out of the HAE UAV ACTD Program*, Santa Monica, Calif.: RAND Corporation, MR-1476, 2002. As of September 19, 2008:
http://www.rand.org/pubs/monograph_reports/MR1476/

"F/A-22 Logistics Privatization Study (LPS)," April 30 , 1995. Not available to the general public.

FAR—*See* Federal Acquisition Regulation.

Federal Acquisition Regulation, Subpart 9903.2, "CAS Program Requirements," Appendix, 2005.

Federal Acquisition Regulation, March 2005 edition. As of March 28, 2008:
http://farsite.hill.af.mil/farsite.html

Gansler, J. S., Under Secretary of Defense (Acquisition and Technology), *Future Readiness*, May 10, 1999.

GAO—*See* Government Accountability Office.[1]

Glendenning, William H., *KC-10 Organic vs Contract Support Cost Study*, Wright-Patterson Air Force Base, Ohio: Aeronautical Systems Division, Directorate of KC-10, Program Control Division, July 15, 1982.

Gilroy, Robert M., Senior Associate Director, National Security and International Affairs Division, General Accounting Office, *Statement before the Subcommittee on Legislation and National Security of the House Government Operations Committee*, March 14, 1984.

Government Accountability Office, *Should Aircraft Depot Maintenance Be In-House or Contracted? Controls and Revised Criteria Needed*, Washington, D.C., FPCD-76-49, October 20, 1976.

————, Strategic Bombers: B-1B Maintenance Problems Impede Its Operations, Washington, D.C., GAO/NSIAD-89-15, October 1988.

————, Financial Audit: Air Force Does Not Effectively Account for Billions of Dollars of Resources, Washington, D.C., GAO AFMD-90-23, February 1990a.

————, *Military Bases: Information on Air Logistics Centers*, Washington, D.C., NSAID-90-287FS, September 1990b.

————, Financial Audit: Aggressive Actions Needed for Air Force to Meet Objectives of the CFO Act, Washington, D.C., GAO/AFMD-92-12, February 1992a.

---

[1]   Formerly known as the General Accounting Office and so identified on the many of the older GAO documents listed here.

————, *Contract Maintenance: Improvement Needed in Air Force Management of Interim Contractor Support*, Washington, D.C., GAO/NSIAD-92-233, August 1992b.

————, *Air Logistics Center Indicators*, Washington, D.C., GAO/NSIAD-93-146R, February 25, 1993.

————, *Air Force Depot Maintenance: Analysis of its Financial Operations*, Washington, D.C., GAO/AIMD/NSIAD-00-38, December 1999.

————, *Depot Maintenance: Future Year Estimates of Public and Private Workloads are Likely to Change*, Washington, D.C., GAO/NSIAD-00-69, March 2000a.

————, *Depot Maintenance: Action Needed to Avoid Ceiling on Contract Maintenance*, Washington, D.C., GAO/NSIAD-00-193, August 2000b.

————, *Defense Logistics: Air Force Lacks Data to Assess Contractor Logistics Support Approaches*, Washington, D.C., GAO-01-618, September 2001a.

————, *Depot Maintenance: Management Attention Required to Further Improve Workload Allocation Data*, Washington, D.C., GAO-02-95, November 2001b.

————, *Depot Maintenance: Change in Reporting Practices and Requirements Could Enhance Congressional Oversight*, Washington, D.C., GAO-03-16, October 2002.

————, *Defense Management: Opportunities to Enhance the Implementation of Performance Based Logistics,* Washington, D.C., GAO-04-715, August 2004.

————, *Depot Maintenance: Persistent Deficiencies Limit Accuracy and Usefulness of DoD's Funding Allocation Data Reported to Congress*, Washington, D.C., GAO-06-88, November 2005.

————, *Weapon Systems: DoD Should Strengthen Policies for Assessing Technical Data Needs to Support Weapon Systems*, Washington, D.C., GAO-06-839, July 2006a.

————, *Depot Maintenance: Actions Needed to Provide More Consistent Funding Allocation Data to Congress*, Washington, D.C., GAO-07-126, November 2006b.

Hannon, David, "Lockheed Martin: Negotiators Inc.," *Purchasing*, February 5, 2004, Vol. 133, No. 2, pp. 27–30.

Heivilin, Donna, Director, Defense Management and NASA Issues, *Depot Maintenance: Issues in Management and Restructuring to Support a Downsized Military*, Washington, D.C.: National Security and Internal Affairs Division, General Accounting Office, GAO/T-NSIAD-93-13, May 6, 1993.

————, *Depot Maintenance: Issues in Allocating Workload Between the Public and Private Sectors*, Washington, D.C.: National Security and Internal Affairs Division, General Accounting Office, GAO/T-NSIAD-94-161, April 12, 1994.

Kingbury, Nancy R., Director, Air Force Issues, *Depot Maintenance: Issues in Management and Restructuring to Support a Downsized Military*, Washington, D.C.: National Security and International Affairs Division, General Accounting Office, GAO/T-NSIAD-92-23, March 26, 1992.

Kutz, Gregory D., Director, Financial Management and Assurance, *DoD Financial Management: Integrated Approach, Accountability, and Incentives Are Keys to Effective Reform*, Washington, D.C.: General Accounting Office, GAO-01-681T, May 8, 2001.

Laseter, Timothy, "Balanced Sourcing the Honda Way," *Strategy+Business*, Fourth Quarter 1998.

Lorell, Mark A., John C. Graser, and Cynthia R. Cook *Price-Based Acquisition: Issues and Challenges for the Defense Department Procurement of Weapon Systems*, Santa Monica, Calif.: RAND Corporation, MG-337-AF, 2005. As of September 19, 2008:
http://www.rand.org/pubs/monographs/MG337/

Multi-Echelon Resource and Logistics Information Network (MERLIN). MERLIN online site decommissioned spring 2008.

Nihiser, Nike, *Technical Data Rights*, AFMC Law Office, Technology Division, Contract Law Division, undated.

Office of the Secretary of Defense, Cost Analysis Improvement Group, *Operating and Support Cost-Estimating Guide*, May 1, 1992.

———, *Report to Congress on Distribution of DoD Depot Maintenance Workload*, Fiscal Years 2005 through 2007, April 2006.

Oliver, Steven A., Alan W. Johnson, Edward D. White III, Marvin A. Arostegui, "Forecasting Readiness: Regression Analysis Techniques," Air Force Journal of Logistics, Vol. 25, No. 3, Fall 2001, pp. 3, 31–42.

OSD CAIG—*See* Office of the Secretary of Defense, Cost Analysis Improvement Group.

Perry, William J., "Acquisition Reform: A Mandate for Change," white paper, Washington, D.C.: Department of Defense, February 9, 1994.

Public Law 109-364, John Warner National Defense Authorization Act for Fiscal Year 2007, Section 802, Additional Requirements Relating to Technical Data Rights, October 17, 2006.

Secretary of the Air Force and Chief of Staff, United States Air Force, "Lean Across the Air Force": Process Development and Improvement, memorandum to all major commanders, Washington, D.C.: Department of the Air Force, November 7, 2005.

Thirtle, Michael R., *The Predator ACTD : A Case Study for Transition Planning to the Formal Acquisition Process*, Santa Monica, Calif.: RAND Corporation,

MR-899-OSD, 1997. As of September 19, 2008:
http://www.rand.org/pubs/monograph_reports/MR899/

U.S. Air Force, *Committee Staff Procurement Backup Book, FY 2007 Budget Estimates, Aircraft Procurement, Air Force*, Vol. 1, February 2006.

U.S. Code, Title 10, Section 2208(j), Working Capital Funds, January 3, 2005.[2]

————, Section 2320, Rights in Technical Data, January 3, 2005.

————, Section 2460, Definition of Depot-Level Maintenance and Repair, January 3, 2005.

————, Section 2462, Contracting for Certain Supplies and Services Required When Cost Is Lower, January 3, 2005.

————, Section 2464, Core Logistics Capabilities, January 3, 2005.

————, Section 2466, Limitations on the Performance of Depot-Level Maintenance of Materiel, January 3, 2005.

————, Section 2469, Contracts to Perform Workloads Previously Performed by Depot-Level Activities of the Department of Defense: Requirement of Competition, January 3, 2005.

————, Section 2470, Depot-Level Activities of the Department of Defense: Authority to Compete for Maintenance and Repair Workloads of Other Federal Agencies, January 3, 2005.

————, Section 2472 Prohibition on Management of Depot Employees by End Strength, January 3, 2005.

————, Section 2474 Centers of Industrial and Technical Excellence: Designation; Public-Private Partnerships, January 3, 2005.

————, Section 2563 Articles and Services of Industrial Facilities: Sale to Persons Outside the Department of Defense, January 3, 2005.

USC—*See* U.S. Code

---

[2]   Date for this and all other referenced sections of the U.S. Code is that found in cite provided by The Office of the Law Revision Counsel of the U.S. House of Representatives.